D1103044

BARCLAY WILLS'

THE DOWNLAND SHEPHERDS

THE SOUTH DOWNS
c 1920

KEY
LAND OVER 200'
LAND OVER 400'

SCALE OF MILES
0   2   4   6   8

Lindfield

HAYWARDS HEATH

St John's Common
Burgess Hill

River Ouse

River Arun

Bury Hill

Amberley
Ashington   *Ashurst*

River Adur

Washington
Kithurst Hill
Highden Hill
Lee Farm   *Chanctonbury Ring*
Harrow Hill   *Steyning*
Blackpatch Hill   *Bramber*
Burpham   *Myrtle*   *Figdon*   *Beeding*
Michel Grove   *Grove*   *Nebcote*   *Cissbury Ring*
ARUNDEL   *Patching*   *Lychpole Farm*
Clapham   *Charmandean*
Highdown Hill   *Durrington*   *Somphing*
Broadwater
Goring
LITTLEHAMPTON   *Angmering*   *WORTHING*   *SHOREHAM*   *HOVE*
On Sea

Fulking

Wolstonbury Hill   *Clayton*
Pyecombe   *Westmeston*   *Plumpton*
Poynings   *Ditchling Beacon*   *Blackcap*
Saddlescombe   *Standean*
Yarncombe Barn   *Mary Farm*
Patcham   *Stanmer*   *Falmer*   *LEWES*
Newmarket Hill   *Kingston*
Bevendean   *Iford Hill*
Rottingdean
BRIGHTON
NEWHAVEN

Ditchling

River Cuckmere

Glynde   *HAILSHAM*
Beddingham
West Firle   *Selmeston*
Firle Beacon   *Alciston*
Wilmington   *Polegate*
Windover   *to Pevensey*
Alfriston   *Hill*   *Longman*
Jevington
West Dean
SEAFORD   *Exceat Farm*
Foxhole
Friston
Cuckmere Haven   *East Dean*
Birling Farm   *EASTBOURNE*
Birling Gap
BEACHY HEAD

BARCLAY WILLS'

THE DOWNLAND SHEPHERDS

Edited by
Shaun Payne and Richard Pailthorpe

With Paintings by
GORDON BENINGFIELD

Foreword by BOB COPPER

ALAN SUTTON
1989

ALAN SUTTON PUBLISHING
BRUNSWICK ROAD · GLOUCESTER

ALAN SUTTON PUBLISHING INC
WOLFEBORO · NEW HAMPSHIRE · USA

First published in this edition 1989

Copyright © this edition, Shaun Payne and Richard Pailthorpe 1989
Copyright © Introduction, Shaun Payne 1989
Copyright © Colour illustrations, Gordon Beningfield 1989

All rights reserved. No part of this publication may be reproduced, stored in a retrieval system, or transmitted, in any form or by any means, electronic, mechanical photocopying, recording or otherwise, without the prior permission of the publishers and copyright holder.

British Library Cataloguing in Publication Data

Wills, Barclay, *1877–1962*
Downland shepherds.
1. East Sussex. West Sussex. South Downs.
Social life, 1910–1936
I. Title II. Payne, Shaun III. Pailthorpe,
Richard
942.2′6083

ISBN 0-86299-408-X

Library of Congress Cataloging in Publication Data applied for

Endpapers: Photographs by Dr. Habberton Lulham (Mansell Collection)

Typesetting and origination by
Alan Sutton Publishing Limited
Printed in Great Britain by
The Guernsey Press Company Limited
Guernsey, Channel Islands

— CONTENTS —

Foreword by Bob Copper ... ix

Introduction ... xi

The Man and the Hills ... 1

First View ... 1
The Better Way ... 4
Two Days of Liberty ... 6
A Stranger in Downland ... 10
The 'Archipeligo Bloke' ... 11
The Wind on the Downs ... 13

The Shepherds of the Downs ... 16

The Lure of the Shepherd's Life ... 16
Some Downland Shepherds ... 20

Nelson Coppard ... 21
Jack Cox ... 23
Michael Blann ... 26
Walter Wooler ... 30
George Humphrey ... 32
Charles Trigwell ... 33
Frank Upton ... 35
William Duly ... 38
George Bailey ... 39
Charles Funnell ... 40

A Ramble with Nelson Coppard ... 42

The Shepherd's Year ... 45

Lambing ... 45
Lantern light ... 45
Lambing time ... 49
A day at a Downland lambing-fold ... 49
Sheep-washing ... 53
Sheep-marking ... 57
Sheep-counting ... 58
Shearing and Shearing Gangs ... 59
Trimming ... 68
Findon Fair ... 69
The Shepherd's Christmas ... 72

Shepherds' Gear ... 73

The Shepherd's Possessions ... 73
Shepherds' Umbrellas ... 78
Horn Lanterns ... 80
Shepherds' Clothes ... 82
The Roller-wattle ... 85
Shepherds' Sundials ... 87
Whistle-pipe and Song book ... 88

Downland Sheep Crooks and their Makers ... 91

Downland Sheep Crooks ... 91
Crook-sticks ... 93
My Favourite Sussex Crook ... 93

A Visit to Falmer Forge   97
Pyecombe Hooks   98

Bells and Bell Music   101

Downland Sheep Bells   101
  Canisters   104
  Cluckets   105
  Wide-mouthed Iron Bells   105
  Lewes Bells   106
  Wide-mouthed Brass Bells   106
  Cup Bells   107
  Horse Bells or 'Latten Bells'   107
  White Latten Bells   107
  Crotals or Rumblers   108
Tackle for Sheep-bells   109
  Wooden Yokes   109
  Leather Yokes   112
  Strap Collars   112
  Chin Boards   112
  Bell Straps   113
  Reeders   113
  Lockyers   113
A Ring of Canisters   114

The Shepherd's Companion   118

A Dewpond Chapter   121

Sheep in Distress   125

The Shepherd's own Page   125
Under the Turf   125
Airplanes on the Downs   127
The Flood   128

Downland Nature Notes   131

Cissbury Ring in June   131
July in Downland   133
Colour in the Cornfield   134
The Wren, and Some Others   135
Dartford Warblers   136
The Stonechat   138
The Wheatear   139
The Salvington Goldfinch   140
The Green Hairstreak Butterfly   140
Clouded Yellows   141

Something of Sussex   144

Sussex Oxen   144
The Hurdle Maker   146
Ashurst Mill   148
If I had Money!   150

The Passing of the Downland Shepherds   152

'This Hidden Sorrow'   152
The 'Furriner'   154
Where Golf Balls Stray   155
The Crow-scarer   158
'I Be 'bout Done'   159

Glossary of Shepherding and Dialect Words   165

Notes   171

Places of Interest to Visit   177

Further Reading   178

Acknowledgements   180

Index   182

Barclay Wills' sketch of a Goldcrest

Barclay Wills, c. 1937. He is holding Frank Upton's crook

— FOREWORD —

by

Bob Copper

It is sometimes possible to look back over the years at an incident which, although seemingly of little significance at the time, turned out to have a profound effect on one's life. This can certainly be said of the day exactly fifty years ago when as a young policeman in Worthing I was called upon to convey Barclay Wills to the General Hospital by ambulance. There was about this man a quiet, dignified charm which together with his steady, grey-blue eyes and patriarchal beard lent him an air of almost scriptural serenity. I felt instinctively attracted to him.

I visited him in hospital several times and, as our acquaintance grew, continued to see him at his home after his return to good health. It was there I discovered his interest in the Sussex downland. He showed me his exquisite drawings of its flora and fauna and photographs of his friends the shepherds, who at one time were to be found tending their flocks on the hillsides. He gave me copies of his three books in which he had recorded so truthfully and simply the kind of world that I had known so well at Rottingdean, only a few miles along the coast.

This was the land where the tang of the sea mingled with the scents of harvest; where channel breezes whistled through the grass-bents and the air was loud with lark song.

These were the hills which for centuries had supported the great flocks of sheep that had made the Southdown breed famous the world over. All my male relatives for generations past had worked as shepherds and farm-hands on the southern slopes of these seaboard hills, so I knew how faithfully Mr. Wills had written about that sturdy and singular race of men – the shepherds of Sussex.

By 1930, as a result of the decline in farming fortunes and much of the land being made over to building development, we were forced to shake the soil from our boots and turn to more urban pursuits as a means of livelihood, and I for one was in danger of forsaking my heritage. But meeting Barclay Wills and being infected by his enthusiasm for and love of everything to do with the downland made me aware of what I would be throwing away. He encouraged me to write and draw and to preserve the past in any other way I could and we formed a firm friendship which lasted until the time of his death.

I like to think of him out on the hills armed with his note-book and sketch-pad, his tall, rangy figure striding up the steep, downland pathways, every sense alert to the sights, sounds and smells that made up the magic under whose spell he had fallen. He was the keenest of observers and would be equally aware of the omnipotent, surrounding mood of nature and of a tiny beetle scrambling through the tangled grasses at his feet. His writings have captured it all.

His 'den' was a mini-museum of 'downland treasure': sheep crooks and bells of all descriptions, sheep shears, dipping hooks and horn drinking cups, which told of his long association with the men of the hills. Each item had its individual history of how it had come into his possession and I listened with rapt attention. He also had an extensive collection of flint implements from earlier ages all discovered on his rambles over the hills – arrowheads, scrapers, hammer-stones and pot-boilers from various '-olithic' periods, all meticulously labelled and placed neatly on cotton wool beds in small cabinets. In that small room where the talk was always about the hills, the shepherds, the tools they used, the tales they told and the songs they sang, I was taken back to my boyhood and all the things I loved.

I am honoured to be asked to write the foreword to this book and feel that in this small coin I am, perhaps, partly repaying the tremendous debt of gratitude which is still outstanding.

In 1938 I wrote the following verse. In 1988 I have no desire to alter a word of it.

AN APPRECIATION OF BARCLAY WILLS

Good friend and champion of a dying race,
The downland shepherds know his genial face
Throughout the length of Sussex. He became
Custodian, in everything but name,
Of all they held most dear. The treasured store
Of sheep-bells, crooks and tales of Sussex lore,
Forgotten customs, ancient song and rhyme
He garnered 'gainst the ravages of Time.
From early shepherd life he raised the mask –
Posterity is richer for his task.

'Tis odd indeed the fact that it should take
A 'furriner' to nudge us wide awake
To all the charms our native acres hold.
By lantern-light he leads us to the fold
At lambing-time, and with him we have seen
The sun behind the mill at Rottingdean.
But though his laugh rings yet on Bury Hill,
It might be said that he's a townsman still.
Then, as one thrust on Sussex, may I voice,
This man became Sussexian from choice.

Bob Copper, 1988

— Introduction —

Six years ago, while reading along the dusty shelves of a secondhand bookshop I chanced upon a fat, blue, hardcovered book with faded lettering on its spine. I took it down and turned to the title page: '*Bypaths in Downland* by Barclay Wills . . . with fifty-eight illustrations from photographs and drawings by the author . . . *First published in 1927.*' I leafed through, noticing first the illustrations and a few chapter headings: photographs of shepherds. . . dewponds. . . LANTERN LIGHT. . . Jenny Wren. . . CISSBURY RING IN JUNE. . . more shepherds. . . a blacksmith. . . NOTES FROM A HEDGEBANK IN MARCH. . . sheep bells. . . a drawing of 'lockyers' (what were they?). . . oxen ploughing. . . THE SHEPHERD'S OWN PAGE. . . I read at random, and was enthralled:

The springy turf of the Downland bottom was pleasant after the meadow. The gorse on the slopes seemed alive with linnets, and as I climbed the steep path to the brow a wheatear acted as a pacemaker, flying off in advance and settling again and again. I did not overtake him, however, for a side track showed wheel marks and certain imprints which were sufficient to direct me, and in a few moments the winding path brought me in view of the shepherd waving a welcome.

My mind filled with questions. Why had I never seen this wonderful book before, nor heard of its author? Had he written anything else? Where had he lived and when? Only afterwards, when I had returned home, did I notice something else about my 'treasure'. There was marginalia on several pages, written in pencil in a clear beautiful hand (which I was later to know as his), and inside the back cover was a plain postcard, which fell out, dropping at my feet. On one side was a typed note, dated '22/12/50':

Perhaps you would like to know that the hand bells which you described in your book, *Downland Treasure*, will be seen and heard on Television Newsreel on Christmas Day at 8.0 p.m. . . . playing carols in Albourne street. . . .

On the other, an address:

Mr. Barclay Wills,
57, Ham Road,
Worthing.

In the seconds that followed I knew I had to try to discover more about him.

I soon found that my ignorance of Barclay Wills' work was not very surprising. He was not mentioned in any of my own books and scholar friends whose work is the study of country writing could not enlighten me either. His other books, *Downland Treasure* (1929) and *Shepherds of Sussex* (1938), had been out of print since first publication and were hard to get. I did discover eventually that Wills' work still had a small but knowledgeable readership, but why, I asked myself, were his books so little known? To read his writings is to enjoy the companionship of an exceptionally observant guide, artist and naturalist, whose knowledge of the life of the shepherds was such that he was rightly regarded as the 'supreme authority'. Yet Barclay Wills was no 'dry as dust' expert: his prose is simple, fresh, and compellingly alive. To read him is to find delight and to share his boundless enthusiasm for the countryside.

Very little is known of Barclay Wills' early life. He was born 'a poor Cockney' on 22 June 1877, at 23, Florence Street, Islington. His full name was registered as 'Harry Barclay Wills', and he seems always to have been Harry to his family. Barclay, which he was to use for his books and pictures, may have been given him in memory of his father's mother, Sophia Wills, whose maiden name is recorded as 'Barklay'. His father, Henry William Wills, and his paternal grandfather, also Henry William, worked in publishing, though for which publisher and in what capacity is unknown. His mother, Lydia, was the daughter of James Hoare, from Limehouse, a clerk in H.M. Customs, and the couple had married at the Church of St. Philip in Clerkenwell seven years before Barclay's birth. On the evidence of those who knew Barclay Wills well in later life it seems likely he was an only child – he certainly never mentioned having a brother or sister.

Despite his London upbringing, Barclay Wills seems to have had from an early age a 'passion for birds and all wild creatures, for trees and flowers, subjects for sketches and old-fashioned things; everything, in fact, connected with the countryside'. A watercolour of a peacock butterfly which he painted when he was thirteen survives, and by 1903, when he was twenty-six, he was producing bird and other pictures of professional quality for book and magazine illustration. Several were published – among them his study of a Marsh Tit in 1906 – but whether he was able to make his living as an artist at this time seems doubtful. In 1905 his occupation was recorded as 'clerk'; three years later, as 'Colonial Importer's Clerk' – jobs equally soul-destroying for a man of his temperament. He was later to remark[1] that he had 'only survived by earning a reputation for eccentricity, for my sketch-book and binoculars saved me from the hopeless futility of suburban life'; and, more bitterly, that he had 'been as a caged bird for so many thousands of precious hours . . . The only use of all my slavery has been to give double keenness to my joy in short spells of freedom.'

These 'short spells of freedom' were spent exploring the leafier fringes of North London, where he lived as a young man – Cow Wood (Highgate), Hampstead Heath, and other quiet spots where he could find wild plants and birds to sketch. He also made other, perhaps longer, visits outside London, especially to Hampshire, the Isle of Wight, and the New Forest, where he experienced 'that wonderful peace which pervades an ancient solitary spot'. In December 1903 he drew a Stonechat while at Bournemouth Chine, and in April and September of 1904 was at Brockenhurst in the heart of the New Forest. Here he sketched King Cups, a White Admiral butterfly, Sallow buds, and made several drawings of the Nuthatch, a bird whose movement, like that of the similar Tree Creeper, always fascinated him. Among other early sketches in the Worthing Library Collection are two that are particularly intriguing: a pencil study

Barclay Wills' early sketch of a 'Wren ascending trunk as a Tree-Creeper does'

of a 'Wren ascending trunk as a Tree Creeper does', which is marked 'Falmer Wood'; and of a 'Cuckoo resting', made at Findon. As neither is dated, it is impossible to be sure just when he may first have visited these places which were to become so important to him, but the dating of other sketches within the collection seems to indicate that it was between 1903 and 1917, the first and last dates which he noted on the drawings.

Most of the two hundred or so natural history sketches and illustrations by 'BW' that have been traced are of birds, but he also drew plants, insects, mice, voles, and his pet dormouse. A few sketches show that he sometimes made careful studies of detail from dead specimens, including on one occasion a Dartford Warbler 'shown to me in a taxidermist's shop'. His usual method, though, was closely to watch and draw birds in the wild. The finished illustrations that were made from these quick, first sketches are outstanding, and are all the more remarkable for being the work of a self-taught artist. He captures the attitude and movement of each bird, and conveys that indefinable softness of certain species, such as the Grey Wagtail, and it is not surprising, in this respect, that he knew the work of Archibald Thorburn. He may, indeed, have known Thorburn himself: among the pictures at Worthing Library is one of a 'Cream-Spot Tiger Moth found in Archibald Thorburn's garden at Godalming'.

Many years were to elapse, however, before he could make good his escape from London, and it was not until after the Great War that he was able to fulfil his youthful dream of living in Sussex, close to the Downs. By then, he was in his early forties; had spent several years in a publishing office (following family precedent); and had married, at the age of twenty-eight, Bertha Paddock (whose father, James, was a builder's foreman), while he was living

To Daphne the opening buds and uncurling fern fronds are a real joy. The baby rabbits, birds, and lambs claim her love. Leaf mould, moss, thyme, and wild fruits thrill us both. With the rising of the sap in the trees we are always seized with a vague longing for the woods and places where our favourites live; therefore we rambled out to-day to a little wood where green woodpeckers called, where anemones, primroses, and violets were waiting for us.

Sadly for him, this mutual enthusiasm seems not to have outlasted Mollie's childhood. Although she lived at home throughout her life and remained devoted to her father, Mollie seems to have had little understanding of his genius, and Bertha, remembered by many as 'a sweet woman', had little insight into the true nature of her husband's dreams.

When Barclay Wills eventually moved from London, probably in 1922, it was not, as he must have hoped, to the Sussex countryside, but to Brighton, where he took a small general shop at 50, Lavender Street, in Kemptown. We do not know what caused him to leave publishing, nor why a man who so hated town life should have gone to live in Brighton, that 'Little London-by-the-Sea', but it is evident from hints in his books and from correspondence that the years following the move were financially difficult ones, and that it may have been, like so much else in his life, a matter of pounds, shillings, and pence.

In the years following the move from London he ran a succession of shops and small businesses. The venture at Lavender Street[2] seems to have been short-lived. By January 1924 the Wills had taken The Willow Cafe, in Dyke Road, Brighton. Again, their stay was brief, as they soon moved to take a grocery business at 59, High Street, Worthing. Here they remained for four years, before

Barclay Wills' drawing of a Cream Spot Tiger Moth, found in Archibald Thorburn's garden at Godalming. Barclay Wills admired Thorburn's work, and may have known him

at 109, North Hill, Highgate. Three years later, he and his wife had moved to Hernside, Western Road, in Fortis Green, where their daughter, Mollie Barclay Wills, was born. She was to be their only child and was given the pet name of Daphne. She was later to share many rambles with her father, and he seems to have taken particular delight in her childish enjoyment of the countryside:

moving again, this time to run another's grocer's shop in the then small village of Durrington, at 5, Salvington Road, near the Lamb Inn. In 1932 there was another change of address: 1, Brougham Terrace, Brougham Road, East Worthing, where the family lived above the shop, also a grocer's. None of these businesses seems to have been successful commercially. A Worthing housewife who shopped at the Wills' 'because I liked them' told me, 'He wasn't good at running the shop – his wife kept things going.' His lack of interest is not surprising: although he valued the independence that working for himself gave, being a grocer can have been hardly more congenial than clerking – it, too, kept him from the things he loved to do best, from his drawing and writing.

The move to Sussex was, in other ways, a happy one; he was near the Downs 'at last'. Worthing and even Brighton were much smaller than they are today, with unsullied countryside close to both. Although, by the mid-1920s, some Downland had been lost to golf courses and suburban housing, large areas remained as yet untouched. The town-tired rambler could still find quiet spots, far from the din of motor traffic, unchanged since Gilbert White rode the Downs; could still watch, in a few places, the slow tread of oxen ploughing; or, resting on a bedstraw-scented brow, listen while sheep bell music floated up from the green bottom below, and lark song flooded the sky. He could still feel as he rambled over mile upon mile of open Downland, the soft, elastic 'spring' of the turf underfoot; could pause to wonder at the minutely magical flowers of the close-cropped sward – flowers such as eyebright, chalk wort, wild thyme, and harebell; or go, field glasses in hand, to one of the tracts of gorse-covered chalk heath, haunt of the Dartford Warbler and that (now) rarest of Downland butterflies, the Silver-studded Blue. Barclay Wills, of course, did all this, and

more. For him, the Downs were an enchanted region – in W.H. Hudson's phrase, 'a fairyland'.

On one of his first Downland rambles, he went to Falmer, four miles from Brighton. There he heard 'for the first time. . . the music of old sheep bells' and met the shepherd, Nelson Coppard. The two men warmed to each other. Soon 'Mus. Wills' was a frequent visitor to Mary Farm, proud to call Nelson 'my friend the shepherd'. The friendship was to prove momentous, as important for him as Sturt's with Bettesworth or Hudson's with James Lawes (Caleb Bawcombe of *A Shepherd's Life*). He had found his subject and a guide to help him. From Nelson he had his 'first instruction on sheep bells, crooks, and the details of a shepherd's life'; through him he was able to meet other shepherds, certain of a friendly welcome. Not surprisingly, when he began planning his third book, *Shepherds of Sussex;* his first step was to discuss it with Nelson.

Nelson Coppard's help in putting Barclay Wills in touch with other hill shepherds was particularly valuable. The Downland shepherds' solitary work made them men apart and, like many countrymen of their generation, they were wary of strangers, especially those who spoke (as Barclay Wills did) with a town accent. Other writers had tried (and failed) to win the trust of these men. As Wills himself pointed out, everything depended 'on the attitude of the shepherds' and it was, he knew, essential to gain their confidence. Where others might have visited an old shepherd once, 'Mus. Wills' went back again and again, listening sympathetically while the old man repeated himself or grumbled, perhaps, about the doings of the new farmer. In time he gained the affection and respect of many of them, including Tom Rusbridge of Findon, George Humphrey of Sompting, and Michael Blann, a retired shepherd at Patching. When he began, in the late 1920s, to research his history of the Sussex shepherds he was enthusiastically helped by the men themselves:

They welcomed the plan of a book devoted to themselves and their craft, and responded willingly to requests for details of their lives and information on many subjects. Family treasures were unearthed; old days were remembered again; names and places were quoted, and I was able to follow many trails which resulted in the gathering of valuable facts regarding former habits that had been almost forgotten under the stress of modern ways in sheep farming.

By the mid-1920s the 'stress of modern ways' was starting to have other effects on Downland agriculture. In 1924 Brown's Farm at Rottingdean was sold. It was to be the first in a series of such sales as the old families who had farmed the hills for generations were forced to sell to ousiders ('furriners'), who understood neither the shepherds nor the Downs. Worse still, the Downs were under increasing threat from speculative builders. In 1921, the South Coast Land and Resort Company had begun to build a new town (Peacehaven) on open Downland to the east of Brighton. It was, many felt, an ominous development and, in December 1923, the Society of Sussex Downsmen was founded to try to protect the Downs. Barclay Wills was among the first to join the Society (in January, 1924), and was to be one of its most ardent supporters. He served on several committees, and some early meetings were held at the Willow Cafe before more permanent arrangements could be made. Later, after his move from Brighton, he became a District Officer, responsible for keeping an eye on a section of the Downs near Worthing. In 1929, when developers had a huge hoarding cut into the turf near Cissbury Ring, he was quick to act, setting out in soaking rain from Durrington to Salvington ridge, from which he got a view of the advertisement. He wrote to Lilian Bately, the Honorary Secretary of the Society, that 'no protest could be too strong regarding the awful effect of the lettering . . .' (Fortunately, the Society was successful in its opposition). More usually he found himself investigating the loss of footpaths and rights of way.

The rapid changes caused by the coming of modern agriculture, urban sprawl, motor traffic, and what C.E.M. Joad[3] called 'the Untutored Townsman's Invasion of the Countryside' threatened not only the beauty and peace of the Downs but also the way of life of those, such as the shepherds, who belonged there. Barclay Wills was profoundly aware of this. He sought to record what he could of their folk ways and rural culture while it was still possible to do so. While in his forties and in indifferent health, he would tramp miles across the hills to meet and photograph old countrymen, afterwards sitting on the open Down to write his notes while their talk was still fresh to memory. He saw the last ox-team ploughing the Downs, went to see the hurdle makers at work in the Sussex woods, and befriended the old village blacksmiths famous for making shepherd's crooks.

Wherever he went in the countryside Barclay Wills collected 'bygones' that would serve as a record for future generations: Sussex pottery, horn-cups, lanterns, pewter, old brass and copper, horse and cattle bells, ox-cues, shepherds' crowns and lucky stones, smocks, and all manner of old farming gear. Pride of his collection was his shepherding tackle, including crooks and sheep bells, which grew to become one of the most important reference collections in the country. Many, perhaps most, were gifts from Nelson Coppard and other shepherd friends, others picked up for a shilling or nothing. To those who complained that 'the one and only place for a sheep bell is on a sheep's neck' he pointed out that he collected them as a tribute to the memory of the men who had given him them. He gave much of his valuable collection to Worthing Museum, where many of the items mentioned in his writing, such as Michael Blann's song book, may still be seen and enjoyed. He worked closely with Marion Frost, the Museum's first curator, and her successor Miss Gerard, giving them detailed information about the history, provenance, and use of each exhibit, even suggesting the wording for exhibition labels.

Through the Society of Sussex Downsmen Barclay Wills was able to meet others who shared his feeling for the Sussex countryside. Among the small group who founded the Society were Dr. Habberton Lulham and Arthur Beckett, who were to become (if they were not already) his close friends and who were to have a profound influence upon him. Lulham, twelve years his senior, was a kindred spirit: a poet, he too had spent days and nights with the shepherds, collected sheep bells, and counted men such as Jim Fowler and Tom Rusbridge among his friends. He was also a fine photographer whose beautiful sepia pictures recorded well the mood of the Downs and the life of the shepherds. When the first of Barclay Wills' articles about the Sussex shepherds was published in March 1924 it carried Lulham's picture of Jim Fowler. Later, Lulham was to contribute a fascinating chapter of 'Stray Memories' and seven pictures to *Shepherds of Sussex* (1933–4). Arthur Beckett – the title of whose book, *The Spirit of the Downs* (1909), had become something of a catch-phrase among Downsmen – had known Stephen Blackmore (1833–1920), the famous shepherd of East Dean. A man of letters, whose library contained some four thousand works relating to

Sussex, Beckett was regarded as *the* authority on the literature, history, and folk life of the county. Like Lulham, he was among those who had assisted Miss Gosset with her important anthology, *Shepherds of Britain* (1911). As the proprietor of several local newspapers, he was a man of some wealth and influence, able to offer Barclay Wills encouragement and practical help. It was Beckett, who when he 'had finished reading Mr. Barclay Wills' first book . . . earnestly advised him to write the history of the Sussex shepherd'. When two years later, Wills had pleaded that he was 'too poor . . . to undertake it', Beckett persisted, publicly pointing out that if he did 'not write the book which he can write better than any other man he will fail to do himself justice'. Happily this advice was heeded and Beckett was able to publish the *Shepherds of Sussex* as a series of monthly articles in the *Sussex County Magazine*, of which he was the Founder Editor, during 1933 and 1934.

Shepherds was not the only book to receive its first publication in Beckett's magazine. Under his inspired editorship, it was rich in original, well-researched articles by contributors such as Lulham, H.S. Toms, J.H. Pull, Maude Robinson, R. Thurston Hopkins, and others Wills knew. Its pages touched upon almost every aspect of the literature, folk lore, and rural life of Sussex. While much of this was of purely local appeal, the magazine also reflected national concern over the future of the countryside. There was tremendous popular interest at the time in the landscape and traditions of the Southern countryside, reflected in the many rural and topographical books of the period, of which H.J. Massingham's *English Downland* (1936) was an example. Of all counties, Sussex was the most written about: where Richard Jefferies (*Wild Life in a Southern County*, 1879; *Nature Near London*, 1883), W.H. Hudson (*Nature in Downland*, 1900), Hilaire Belloc (*The Four*

'The Shepherd's Companion' was one of a series of articles first published by The Sussex County Magazine; it appeared in the May 1934 issue

317

THE SHEPHERD'S COMPANION

to wait for t' other to help, and not *try* to work alone."

It is the careful, patient training of dogs which is responsible for the wonderful work which surprises us. Every shepherd has his own ways, and his dog is used to them, consequently we may see dogs obey signals made by a raised arm, or a motion with a crook, or respond to nods of the shepherd's head, movements of his eyes, whistles, calls, muttered words, and other half-secret signs.

The following incident was related to me by a friend who had been studying the ways of shepherds in connection with a book which he was writing:

"While there was a mist on the hills I went to find Shepherd G., just for the experience. I wanted to hear unseen sheep bells and to study the details of a shepherd's work at such a time. After much difficulty I located the flock and G's. collie found me. As you know, he is not a very safe animal. He rushed at me when I appeared, but fortunately did no harm. When I met G. I told him that I had had difficulty in finding him and that I wished to accompany him in his walk through the mist. To my surprise he asked me

Men, 1912), and Rudyard Kipling (*Puck of Pook's Hill*, 1906) had gone, many others followed. In the thirty years from the first edition of *The Spirit of the Downs* (1909) several hundred works were devoted to Sussex. While most of these were ephemeral, a few have real merit as records of a vanishing way of life and still read well today, including R. Thurston Hopkins' *Sussex Pilgrimages* (1927) for which Wills wrote an appendix on shepherding.

Bypaths in Downland (1927) and *Downland Treasure* (1929) were published by Methuen, whose chairman E.V. Lucas was deeply interested in Sussex. Like many other rural books, both were a gathering of smaller items – poems, anecdotes, and short articles, illustrated by Barclay Wills' own sketches and photographs. Both books told of his adventures among the hills in search of 'Downland treasure', a term so personal that it could happily apply to the discovery of a long sought after sheep bell, the shed antler of a fallow deer, violets from a Sussex wood, or a picture taken at a lambing fold with his box Brownie. Both books, of course, abound in portraits (in words and image) of shepherds he knew, and in records of his talks with them. Invaluable as these are *as records*, Barclay Wills never gives the impression, as even Richard Jefferies sometimes did, of 'emptying his notebook'. There is a lightness of touch about his writing that raises it above the level of mere factual reportage. Although it must be said that the two books are are uneven in achievement – *Treasure*, especially – there is about them, as one reviewer put it, 'a peculiar enchantment' that is strangely resistent to critical analysis, but which seems to flow from Wills' personality and from his fascination with almost all he found in Downland. This point was well made by the reviewer for *The Nation and Athenaeum*: '. . . free from all conscious straining after eloquence, the quietness and simplicity of its language go further than elaborate descriptions in conveying the peace of Downland.'

Bypaths and *Treasure* were widely and favourably reviewed in the national as well as the local press. By 1938, however, when *Shepherds of Sussex* was finally brought out in book form, the public had begun to weary of books about Sussex, a point tacitly acknowledged in its 'Author's note': 'So many books about Sussex have been written that lovers of the county must wonder whether there is room for another.' To make matters worse, *Shepherds* rather unluckily came out at the same time as *Shepherd's Country*, by the much better known H.J. Massingham. As a result reviews at national level were few and those, such as *The Daily Telegraph*, which did review it did so in a cursory fashion, while devoting much space to Massingham's somewhat misleadingly titled book. It was left to Beckett, writing in the *Sussex County Magazine*, to acclaim Barclay Wills' calling the book a 'Sussex classic'. But it is, I believe, much more than this; it is quite simply the most authoritative work in the literature of shepherding and one of the finest rural books of our century. In it Wills showed an understanding of the shepherds and a knowledge of their craft that has been equalled by no other English writer.

Another aspect of Barclay Wills' passion for the Downs was his deep interest in the prehistorical archaeology of the area, which led some shepherds teasingly to refer to him as the 'Archipeligo bloke'. However, it was to a shepherd friend who had shown him how to spot the worked stones, such as arrowheads, axes, scrapers, hammerstones, knives, and the like, that he was indebted for his first interest in flints. Such a hobby fitted in well with his other Downland interests, especially as the best time for searching many spots was in the winter when the fields were bare and wildlife less to be seen. Once home, he would wash and scrub a flint clean and then sit, often in the dark, or with eyes closed, while he turned it over in his hand until he knew how it had been held and hence its use. He kept notes and made many drawings of his finds, recording the place and date of each. As an artist, he was particularly taken with the subtle, varied colour of flint:

I have a box full of flint 'scrapers', selected from the hundreds I have found, the chief charm of which is their colour. Their delicate shade fascinate me. In them I find repeated tints of earth and sky; greys, blues, and browns, relieved by specks and lines of rust colour and other markings; spotted flints, dotted flints, banded flints – no artificial effects, these, but beautiful pieces of colour picked out of the furrows.

As his interest grew, he made contact with other flint-finders and archaeologists, including John H. Pull of Worthing, whom he had met by 1925, and who was to be a close friend until his death in 1960. A postman, John Pull, was also a gifted amateur archaeologist, who, working with his friend C.E. Sainsbury, had begun to excavate Neolithic flint mines at Blackpatch, artefacts from which have recently been dated to *c*. 4200 BC. In *The Flint Miners of Blackpatch*, Pull acknowledged Barclay Wills' help and attested to his skill in spotting important finds among surface debris at the entrance of the mines. During 1932, 1933 and 1934, he helped Pull and Sainsbury with the excavations of other mines at Church Hill, Findon. Under Pull's influence, he himself became an authority on prehistory and his collection of flints was so highly regarded as to be one of only six private collections to be listed by Dr. Curwen in his *Archaeology of Sussex* (1954). This, along with his many other collections of 'Downland treasure', was kept in his crowded study, known to friends as his 'den'. He was, as he later said of himself, 'an incurable collector'!

During the years of the Second World War Barclay Wills lived quietly with his family at 57, Ham Road, East Worthing, to which they had moved a few months before hostilities began. He and Bertha did their best to keep the shop going, but it cannot have been easy. In an undated letter (written in the spring of 1940), he told Lilian Bately: 'We are going on as usual but business is still bad & very difficult.' He also told her: 'I had my first ramble to Cissbury a fortnight ago . . . I just pottered about on the West Rampart . . . I have not been all round the Ring to inspect the place properly for 18 months & still cannot tramp like I used to do.' The last comment was a reference to a chest ailment which had put him in hospital in 1938, and which led to his meeting Bob Copper, then a young policeman in Worthing, and their subsequent friendship.

After the publication of *Shepherds*, Barclay Wills had decided against writing another Downland book, and Habberton Lulham and Arthur Beckett, who might have persuaded him otherwise, had died during the War. Instead, he devoted his precious leisure hours to his hobbies, to hunting for bygones, and to archaeology. With the Downs no longer closed by the military (as they had been for most of the War), he was free to go 'flinting' and to ramble as of old. He again helped John Pull in his excavations, such as that of the flint mines at Church Hill, Findon, in 1946, 1947, and 1948, but he was at the same time keenly pursuing his own research into the flint tools, known as 'eoliths', used by man at a much earlier date. In the four years following the war he searched for these almost daily on East Worthing Beach, as well as at Cissbury, Church Hill, Mount Carvey, and other Downland sites.

Although he remained cheerful and seems to have enjoyed a state of mind which was often close to serenity, the last years were not easy for Barclay Wills. He continued to suffer the bronchitis that had dogged him for years. Money difficulties persisted and he was obliged to dispose of much of his superb collection of bygones, partly to raise cash and partly to find good homes for the most treasured items, as neither his wife nor his daughter took much interest in the crowded contents of his 'den'. A note of the time, hidden in a gift to Charlie Yeates, reflects rather touchingly his sadness that his many interests met with little enthusiasm at home:

I cannot find the other can so it must have disappeared when we moved (like a few more things) – My daughter does not want this one as she is not keen on curios or antiques, except old glass, so it may find a new home now with somebody who appreciates it.

Another disappointment was his failure to find a publisher for his fourth and last book, *Discoveries in Downland*, written during the mid-1950s. This was 'an account of quiet and patient search' on the Downs and the

— INTRODUCTION —

beach for 'eoliths'. These, he felt, had the greatest 'human appeal of anything in any museum' as their discovery linked our own lives with those of 'the first men'. Although his claims for eoliths won support from knowledgeable friends like John Pull and other 'amateurs', professional archaeologists and museum officials were sceptical. Among museum curators only his old friend, H.S. Toms of Brighton Museum, seems to have given him a sympathetic hearing, while the Keeper of one national archaeological collection seems to have been particularly withering, provoking him to write:

I would rather remain a poor flint-finder and enjoy the delight of discoveries in the open air than be associated with sceptics who have explained fantastic theories to me and treated me as I still read 'Jack and the Beanstalk' for amusement!

Discoveries also records his dismay at what he saw as an increasingly materialistic society and his anger at the mindless destruction of so much Downland. He lamented:

When we ramble we meet hundreds of metal pylons, built to carry miles of cables which could have been laid underground. These defile the beauty of the countryside.
'Agricultural Committees', – part of the Ministry of Agriculture, – are now spoiling our downland, without improving our meals, and abusing the power given to them. Large areas are enclosed with barbed wire. Old footpaths and tracks are ploughed in. These vandals even destroyed 'Blackpatch' in 1950, – a famous Sussex flint-mine area which was supposed to be preserved as an 'ancient monument'. After great labour and expense they grew barley on it. This birthright of the public, which has

remained untouched for thousands of years, (as proved by the flint tools and mines and graves discovered there) cannot ever be replaced. It is obliterated. . . .

The loss of Blackpatch must have been especially bitter. A quarter of a century before he had helped John Pull excavate the mines which were among the most important of archaeological sites in England, and for years he had spent 'joyous hours' there in quest of flints and birds. On later rambles to Blackpatch he must have felt a deep loneliness. As was said of one of the old shepherds he knew, 'He was a survivor from a world that had gone.' He did not quite belong any more.

All was not sorrow, however. He might be 'well past the allotted span of three score and ten', but on fine days he could still enjoy a tramp on those parts of the Downs that had, as yet, escaped agricultural 'improvement'. He could still delight in wild birds and creatures. In 1950, while wandering on Mount Carvey he was thrilled when:

a brilliant spot of light. . . shone like a flame in a tuft of dried grass. I crept near it and found that the effect came from the fore-wings of a Small Copper butterfly settled on a stem, in the brilliant sunlight. Then I realised I was staring at a rarity, for there were none of the usual black markings on those wings. They were quite plain, like polished copper. It was a 'variety' of this common little butterfly . . . The metallic effect in sunshine was a delight as I watched it for a moment, then the tiny flame danced away over the field and disappeared.

Nor had he lost his sense of fun; his advice to other flint-hunters on how to dress is typical:

If you worry about your appearance or what other people think of your ways you will not be a good flint-finder or enjoy your hobby as you should do, for even at this date enthusiastic collectors cannot help noticing the strange effect on a few people when *flints* are mentioned. . . . In the past artists, writers and naturalists were always considered to be 'rather unusual' but if you happened to be all these and a flint-finder too you acquired a reputation for peculiar ways. Mrs. A would chat to Mrs. B: – 'There goes Mr. Wills, – I wonder where he is going this time! – poking about somewhere, I suppose!' . . . – 'Oh yes, dear, – a very quiet man, and quite *respectable* I think, but *odd*, – writes and paints and brings home wild plants and things, and collects *old stones*, and *that* really is the limit! . . . It is not our business dear, I know, but people like that – . . .!'

Before their chat was finished Wills was on his way to search for more treasures. Not for the first time 'a reputation of eccentricity' had saved him!

This passage suggests that Barclay Wills must, at times, have cut rather an odd figure. It would be wrong, however, to assume that he was a loner. With strangers he was often shy and reserved, but with those who had the sensitivity to understand him marvellously otherwise. He was a gifted story-teller, who could delight with tales of his adventures among birds and shepherds. He spoke as he wrote, with a soft, strangely magical voice, and could capture effortlessly the cadence and dialect speech of shepherds he knew. His view of life, revealed in these anecdotes, was poetic; his enthusiasm for the old Downland traditions boundless. He had, too, that rarest of gifts – a way of inspiring others to find themselves.

One friend who cheered his old age was the actress and author, Lilian Nancy Bache Price (1880–1970). They had first met during the 1930s, when Miss Price called, apparently by chance, at the Wills' shop in Brougham Road, her curiosity aroused, perhaps, by the exhibits of 'Downland treasure' he was apt to display among the groceries in the shop window. The rich friendship which grew out of that first meeting may have seemed improbable to those who knew Miss Price, star of the London stage, only in her public persona, but there was another side to her personality – that of the recluse, who loved wild nature, befriended birds, and quoted Wordsworth. Besides, Miss Price had a marvellous gift for putting others at their ease. The confidante of Queen Mary and friend of such theatrical celebrities as Beerbohm Tree, she was equally herself with the shepherds, gipsies, and rural 'wayfarers' she befriended. She would have sensed at once the charm that lay behind Barclay Wills' reserve. Moreover, the two had much in common: birdlore, archaeology, the Downs, and an empathy for the old country ways.

The friendship seems to have grown quickly. Barclay was soon a frequent visitor to her cottage, 'Arcana', on the Downs near High Salvington Mill. He helped her add to its decoration, providing her with many of his own photographs and drawings, and fixing up a ring of sheep bells which served her as a door bell. She, in turn, recognised his gifts as an artist, naturalist, and historian. In 1963, a year after his death, she told a reporter how they had shared Downland walks while discussing things of mutual interest. 'His knowledge of such things as pre-historic flints, sheep, horse and bullock bells, plants, birds and the ways and mannerisms of old Sussex shepherds was quite extraordinary.' In another interview she commented, 'He actually knew the shepherds by the bells on the sheep's neck.'

The friendship with Nancy Price was to become one of

the closest of his life. It was she who, after his death, was virtually alone in proclaiming his genius, and it was her efforts which eventually led to a small exhibition being held in his memory at Worthing Museum, in May 1963. At the time she clearly felt that more might have been done by Worthing to honour his memory, remarking to the local press: 'I'm not going to pretend that this show. . . reflects anything like the true genius of the man or the extent of that genius. . . what I wanted to see was a much larger exhibition aimed at recognising his true worth as an artist, naturalist and prose poet.' Whether 'more could have been done' then is open to question. Mr. L.M. Bickerton, the Curator and a friend and ally of John Pull, was strongly supportive of Miss Price's case. In a memorandum, written at the time, he referred to Barclay Wills as 'one of the most learned and endearing men Worthing has known'.

Barclay Wills was eighty-four years old when he died in 1962, apparently of heart failure, at 4, Farncombe Road, Worthing. The date, 1 April, would surely have amused him. He was cremated on 5 April at Woodvale, near Brighton, and his ashes scattered in the Garden of Rest there. His daughter, Mollie, made the funeral arrangements, but there is no record of who else attended the service. Mollie herself died suddenly in May 1970, a year after her mother.

'Barclay Wills lives on in every bird, every flower, and every blade of grass.' – Nancy Price

SHAUN PAYNE
Findon, 1988

A flock of sheep by 'Jill', a post mill on Duncton Down at Clayton, near Ditchling. It was moved to its present site from Brighton by teams of oxen and horses during the mid-nineteenth century

— THE MAN AND THE HILLS —

First View

My first impression of Sussex was a very agreeable one. A friend, with whom I stayed at Frant, took me to a lane off the Wadhurst road in the course of a ramble. At one point he suddenly ascended a bank and beckoned me to a gap in the hedge. 'Look through there,' he said. I did so and was enraptured with the wonderful view. I could not speak to him, for I was spellbound at the sight of the many spires and windmills.[1]

My friend noted my delight, which well rewarded him for his trouble. With appropriate gesture he introduced me, as it were, to the county, by exclaiming, 'That, my boy, is Sussex! There is enough stuff there to last you all your life.' 'Enough stuff' covered my many interests, which were known to him – my passion for birds and all wild creatures, for trees and flowers, subjects for sketches and old-fashioned things; everything, in fact, connected with the countryside.

As he spoke a Pale Clouded Yellow butterfly alighted on a ground-ivy blossom at my feet. The winged treasure came straight to me from the south. It was a good omen, for

although, at that time, nothing seemed more remote than the possibility of being able to ramble about in the fairyland which lay beyond the hedge, yet, by a turn of fortune's wheel, I settled at last near the South Downs, in Sussex-by-the-Sea.[2]

It was a good fairy who guided me to Falmer on one of my first rambles in Downland. There, in a little valley by Mary Farm, I walked through a flock of sheep and lambs. For the first time in my life I heard the music of old sheep bells, and was fascinated by it. Here was something of which I was quite ignorant! I burned to know more and to hear the answers to queries that instantly occurred to me.

By a most fortunate chance the shepherd proved to be Mr. Nelson Coppard,[3] one of the old school, and as he came down the track toward me, with dog at heel, his crook, bright as silver, glittering in the afternoon sunshine, I felt instinctively that I should meet with a kind reception.

I was not disappointed. His gentle courtesy charmed me, and our acquaintance has ripened until I am proud to be able to call him 'my friend the shepherd.' From him I learned my first lessons on bells and crooks, and many other things.

In my endeavour to trace certain further details I was surprised to find what a meagre amount of information was available in print. There were references to shepherds in

Flock by Cissbury with the sound of sheep bells on the Downland breeze

'Before me grazed a flock of sheep and lambs, moving slowly towards me, old and young calling to each other in a bewildering variety of tones. Mixed with their voices was wafted the sound of their ancient bells, and these formed my orchestra – the same soft and pleasing one which had played for a hundred years or more.'

plenty, yet not one book appeared to give the details of all the varieties of sheep bells in use. I therefore set myself the pleasant task of acquiring this knowledge by personal observation. Armed with a photo of Nelson Coppard, which proved a magic passport, I gained entrance to the little kingdoms of many shepherds, where I spent some wonderful hours.

My notebook filled rapidly, for in the course of my rambles fresh bypaths constantly offered new attractions, and at last I decided to gather together the records of some of the beautiful and romantic and interesting subjects found in Downland. The incidents are simply told, for the gatherer of sweet memories from Sussex hills and valleys has no need to draw on his imagination.

Whatever the object of our ramble may be, we always find a thousand other interests by the way, and our biggest 'finds' are only pickings from the inexhaustible store of treasure which awaits every rambler on the South Downs.

Our measure of enjoyment is large or small according to the eagerness of our quest, but it is certain that sometimes we shall step near some of the little things that the fairies have to show us, and the more we discover the more we shall be lured away by the magic spell which haunts the bypaths of Downland.

The Better Way

There are two ways of enjoying a Downland ramble. The first (and most usual) way is to plan out a route, and, while making due allowance for unexpected deviations, to follow a fairly definite track and to visit all the points of interest along the route. Then you are able to say, 'Yes, I have done that walk,' and to compare notes with other people with mutual satisfaction.

There is another way. Forget, if you can, all your maps, guide-books, historical and archaeological treatises.[4] Remember instead some little Downland path which you could not stop to explore at the time you passed it. You will then have a good starting-point for your ramble and you will be your own topographer.

Your outfit must be simple – a weather-proof coat, old clothes, old hat, easy boots, and a stout stick. Minor necessaries should include notebook and pencil, knife, string, matches, needle and thread, and a hairpin. (A hairpin may perhaps seem a somewhat curious item in a man's outfit, but once I found a linnet caught by the foot in a strand of wool which was woven into its nest. I watched the unfortunate bird deftly freed with the aid of a hairpin, and from that time have made use of this wonderful combination tool for many purposes.)

With these items and some food, and sufficient cash to cover the cost of a meal, or conveyance home, as may be necessary, you can start without a care in the world. Do not carry a watch – leave it with the maps and the guide-books, for if you are to enjoy what is offered to the fullest extent you will eat when hungry, recline when tired, and plod when necessary. You will become as a bird released from captivity.

It will be a joyous time for you when you start along your way in this fashion, especially if you tread upon that soft, elastic turf which gives a wonderful buoyant feeling to the body at every step. Your independence of time and destination leaves you free to turn aside from the path at any

'My pet Dormouse'. Barclay Wills' sketch of his pet

point and on the most trivial pretext, or even to retrace your steps and go in another direction as fancy leads you.

It is this sense of freedom which so often lures me into Downland. Through the chase of bird and butterfly the direction of my ramble constantly changes. The flowers delay me all the way. At the sight of a rare or beautiful plant I cannot pass on. I like to sit beside it and enjoy its company, to watch it until I know that the memory of it will never leave me. Thus the hours pass unheeded! Who could hurry away from the shade of a tall hazel bush when a pretty dormouse is making repeated journeys to its hole with its mouth stuffed full of dried leaves? Who could rise from his knees when a nightingale chooses to perch on a twig just above and pour out a rapturous song as if for his special benefit? Who could pass by while a Bee-Hawk moth hovers in the air on tireless wings in front of a bugloss bell? The words 'dormouse,' 'nightingale,' and 'Bee-Hawk moth' take on a different meaning after such intimate association, and the mere sound of them, for ever afterwards, is pleasing to the ear.

Two Days of Liberty

To the rambler it is a fine thing to be free for a whole day, but to be free for two whole days is better still. To know that you can tramp and linger and potter all day as you wish, without any haunting thoughts of the last bus or the last train adds to the joy of your ramble as does spice to the flavour of a cake.

The train from Worthing to Brighton seemed even slower than usual, although a halt by Shoreham Bridge was enjoyable. Plenty of us love to see the view, to look towards Lancing College when the tide is full. The water and reflections, the expanse, the general atmosphere of the scene make an artist wish for a seat on the railway bridge. There is always something to watch. Sometimes one may look down on a heron standing in the shallow creek near the bank, or, when the tide is out on a thousand gulls resting on the mud. The view always compensates one for the rest of the journey along the untidy bit of coastline.

On this occasion I had a pleasant surprise at one point, for as I passed I caught a glimpse of a resting kingfisher. Seated with his back towards me only a streak of blue attracted my attention, but in my mind I saw the bird as I know him, complete with gorgeous coat of blue and green, white, buff, and chestnut, with bright, deeply set eye like a large black bead. In a moment I was borne away from my favourite, but that glimpse of him was sufficient to set my heart beating strongly. The meeting was a good omen and I was anxious to get to Seaford and begin my promised ramble.

A pleasant hour's flinting[5] in a field by Seaford Cliffs was a change from riding, and after two good 'finds' I was more joyous than the gulls appeared to be, for their incessant plaintive wailing screams filled the place. Yet the noisy, wheeling crowd, breaking the silence of the solitary spot, somehow suggested space, and reminded me of the freedom that was mine as I wandered away.

I rested at Exceat, and lingered again to look at West Dean farm, nestling so snugly below me; then on again to Fox Hole, where I expected to find Mr. Dick Fowler,[6] the shepherd. My programme was altered, however, for I found that he had moved to East Dean, and as I thought to see the

'Kingfisher'. Another of Barclay Wills' natural history sketches

ox team there I rambled that way. It was no hardship, for a green track by the roadside was full of interest. Here grew many spikes of clustered bell-flowers and delightful chicory plants in bloom, and they made the way to Friston an easy one. A wandering Clouded Yellow butterfly gave a pleasing touch of colour to the path as it flitted along.

It was here that I heard an unusual word from an old man who sat on the bank. 'I got out o' the las' bus,' he said, 'for a quiet walk, but I allow I'll be safer in it after all. With all they cars swishin' by I got fair scarrified! My son, now, he walks 'bout 's if the cars warn't there. Says he'd like to go flyin' too! "Yes," I says to him, "You're a fool! for there's plenty o' ol'-fashioned ways o' dyin' if you *want* to die, my boy, wi'out tryin' to be a sparrer!"'

By the roadside at East Dean a forge with an old ox yoke above the door offered some interest, but I found that the team of oxen at Birling Farm[7] had gone on a journey from which there is no return. My visit to Mr. Fowler, however, was worth much to me. He owns one of the best lots of canister bells[8] to be heard on the downs to-day and these were in his hut. Most of them are very old and some are only fitted with crown rings of wire. Some belonged to his grandfather and are treasured possessions. He had added to the remainder of the original 'ring' and now has about thirty, including two which are the largest from two other rings. He complained of the difficulty in finding a smith who could be trusted to repair these old bells, which are valued so highly.

In the hut were three wide-mouthed brass bells, and a chance remark about them elicited the fact that these, and some others on his lambs, included those I had heard years ago on another farm. So the rambler on the down links up such items, and gradually finds out what he wishes to know. We went to see the lambs and listened to the merry songs of their bells – the brass ones, a few small horse bells and a latten bell – and the effect of this combination was very pleasing. Mr. Fowler carried one of his Kingston crooks. His best crook is greatly prized and jealously guarded at home. It is only brought out on special occasions.

I left the shepherd at the edge of the farm by the cliff and continued my ramble round Beachy Head. A few gulls flew over from the sea, a wheatear skimmed over the short turf, and a sparrowhawk left me a feather as a souvenir. I envied the wing-power of a tortoise-shell butterfly which danced over the treacherous cliff edge and came back again, while I only trudged along and mingled with the crowd returning to Eastbourne after their lazy afternoon saunter.

There was no room in the Eastbourne hotels for a tired rambler with only a satchel and staff for luggage. One must travel in a motor in these days to get much attention. After an inspection of the museum I drifted to the station. 'Single to Polegate, please,' I said, and was soon out of the town. From the train I noticed a white tower mill, and determined to visit it.

The inhabitants of Polegate are more kind than the hotel clerks of Eastbourne. They do not look past you to note the size of your car; they do not sniff disdainfully when you ask, quite civilly, for a lodging, or look at you with eyes as hard as glass marbles. I was soon fixed up with all the comforts I required and was allowed to make an early start on the following day.

By half-past eight I was tramping again, past flowers and herbage and roadside hedges drenched with dew, and came to the white mill by Stone Cross.[9]

On I went to Pevensey Castle, with its wonderful Roman walls and its ruins of a medieval castle inside. Great progress is being made among the various portions of the ruins, and most fascinating portions are already available for study. As the work proceeds the charm of the place will increase, and when some of the steps and recesses have been scraped and cleaned and scrubbed it is possible that they may prove as attractive as portions of the Old Sarum ruins, where a wonderful effect is seen in the wear of the stones – a mysterious softness, produced by constant human touches, which still lingers there.

Old castles have a great attraction for most of us because they appeal to our imagination, but in one way they are uncomfortable subjects, for their dungeons and other dreadful places still record barbarous customs of medieval days. Such haunting thoughts are not good holiday fare. I turned back again to find another shepherd, for Sussex shepherds are men of peace. I found one in a little hollow between the hills. He and his flock were so well hidden from view as I approached that until I got within fifty yards of the flock there was no sign of life about; then, in a moment sheep and shepherd were in full view.

The shepherd's crook was so bright that I remarked upon it to him. ''Tis said to be the sign of an idle shepherd,' he remarked, 'but stan's to reason that can't be right, for 'tis most industrious I'm sure.' The crook had been broken and brazed, and since it was returned from the forge he had kept it polished. He produced a little tin from his pocket which contained scraps of well-used emery paper. ''Tis near worn out,' he said, 'an' just right for the job, just a little rub every day, an' there it be!'

On I went again, over another hill, down another valley, over a stile and along a stream. I saw tall reeds with plumed tops and long green fingers pointing to a tangle of iris and meadowsweet and other plants; a white-throat searching a bramble bush and a chain of orange and red bryony berries. Then a Downland track led to another windmill, another flock and another shepherd. All these and more made an

'Off to the Hills'. This photograph, taken by Barclay Wills' friend Habberton Lulham, was first published in 1911. The shepherd is Jim Fowler of Ditchling. He is carrying a frail (straw basket)

endless chain of interest until a chiming clock surprised me with a warning that my second day was drawing to a close, and the next bus that passed a certain stile bore me away.

As usual, I took back a few souvenirs of the ramble, but my heaviest item followed me home – a fine canister bell from East Dean. Now as it hangs with the rest, and I note its worn rim and staples, and the lumps of brass which decorate its face, I remember my holiday; I see the shepherd sitting in his hut surrounded by his bells, I see once more a yellow-banded snail-shell on a blue chicory flower, the smoke from couch grass fires on a hill, a quaint iron bucket by a cottage well with a butterfly resting on it and many other fascinating items which are hardly worth recording, yet which filled the hours with sweetness during those two days of liberty.

A Stranger in Downland

From Brighton racecourse I trod the little path to Bevendean Farm which nestles in a hollow at the bottom of the steep hill. Here I met Mr. Norris, the shepherd, with his flock of about two hundred sheep and lambs. He directed me along a steep path leading from the farm to the Downs of Falmer and Kingston.

The cornfield by the path was alive with butterflies to such an extent that I stayed to watch them. The Blues[10] were most numerous and I tried the experiment of feeding one with a raisin. Twenty times I placed the bait gently in front of a settled butterfly; twenty times it was refused.

Then at last a female Blue accepted my gift, stood on the raisin and plunged her little trunk into the sweet, sticky mass. She evidently enjoyed the meal and stood there so long that my finger and thumb were cramped. She refused to move when touched, so I set her down on the raisin among the cornstalks. Once more I had tamed a butterfly in the open! – a trivial triumph perhaps, but better to me than killing one.

At the top of the slope several turtle doves were seen – beautiful summer visitors these, that often rise from the fields at our approach and show us little more than their spread tails, like fans edged with white. Butterflies swarmed here; Meadow Browns, Gatekeepers, Small Heaths, Blues and Coppers flew by in quick succession. Silver Y moths fluttering about amused me for some time, and then I saw, in the distance, a great mass of scarlet – a field of poppies with the sunlight on it. Scarcely had I begun to hasten towards it when I became aware of dozens of white blossoms at my feet. For the first time I had found Burnet roses. In my joy I forgot the poppies and the Silver Y moths, for patches of the tiny, thorny bushes were very plentiful. As I found subsequently it was well worth a later visit to see them in their full beauty.

At last I reached the edge of the poppy-field, where I also found wild heartsease growing plentifully, and it was here, as I rose from my knees after watching a turtle dove through the binoculars, that I saw The Stranger.

'Excuse me,' he said, 'but would you mind telling me the name of that beautiful bird? A turtle dove, eh? Thank you. I must remember that. I really came after you to be directed to Brighton,' he continued, 'but I saw you watching the bird and I became interested as well.' I pointed out his path and told him not to miss the Burnet roses, for his almost childish delight at the sight of the turtle dove, the poppy-field, the heartsease blossoms, the Silver Y moths, the view, a passing cuckoo, and, in fact, everything around gripped me. For once I could admit my own enthusiasm for them all without reserve, although, in frank admission, The Stranger beat me.

'I don't know whether you realize how lucky you are to live in Downland,' he said. 'A doctor has sent me here to roam about for a month, and how I shall go back to the usual daily round I don't know. You don't understand, perhaps, what it means to live in a London suburb; to go to town every day and do the same old job; to come home to a house which you only recognize by the number of it; to do as the neighbours do, and take notice of what Mrs. A. says and what Mr. B. thinks! Oh, it's positively awful, and it will be worse after this freedom! I don't think I have ever been so free before,' he rattled on. 'I roam about as I like and look at thousands of things, although I don't know their names. Here I can lie down if I wish by the side of the path. Here I can eat an apple and pitch the core away anywhere. In our suburb it is *infra dig.* to eat an apple in the street. Now I eat my meals by the roadside for the pleasure of doing so. I suppose it all seems a bit mad to you, but you can't understand how I feel about it all.'

'Pardon me,' I said, 'but I *can* understand, for I am only a poor Cockney myself,[11] and I have been through it all. I only survived by earning a reputation for eccentricity, for my sketch-book and binoculars saved me from the hopeless futility of suburban life.'

I told The Stranger of Patcham Mill,[12] of the steep road above Poynings, of Falmer where the fairies live, of Ditchling Beacon, Lancing Clump, Cissbury, Wolstonbury, and Lewes, of Cuckmere Valley, of bells and oxen and other things. He noted down the names in his pocket-book and at the foot of the page he wrote 'Turtle dove' and 'Burnet rose.'

The 'Archipeligo Bloke'

I was walking over the Downs with a shepherd who had folded his flock and was on his way home. He pointed to some Round Headed Rampion flowers, and said: 'They be pretty! – there be three kinds of this sort of flower, all much alike, as you might say, and one of them is my mother's favourite, but it beant this one, an' I haven't seen many this year – reckon 'tis too dry for 'em or somethin'.' I guessed the plants he referred to and kept a sharp watch as we threaded our way between the many furze bushes. Presently I was lucky enough to find what I looked for – a small Scabious, and a Devil's Bit Scabious, both in bloom. Instantly he picked out the Devil's Bit as the favourite, and he carefully packed away each of the plants in his dinner-bag so that he could take them home and tell his mother their names.

As we passed a field of young corn at the end of the Down a corncrake was calling. It appeared for a moment at the edge of the path, but hid again at our approach. Presently it called after us, and the shepherd remarked: 'Do you hear what he be tellin' us? He keeps on a-sayin' "'tis goin' t' be wet," "'tis goin' t' be wet," – an' most times 'tis true, so don't stay long 'bout or mebbe you'll get wet jacket!' We parted at the cross-roads, and the rain came before I reached home!

Pleasant memories often follow the casual mention of some particular thing. As I wrote the last paragraph I was reminded of another corncrake and hours spent with another shepherd.

In the course of our ramble we heard the call of a corncrake quite near us, among some herbage and low bushes. I told my friend that although I had caught many glimpses of these birds at various times I had never seen one in flight. 'Ye might see it now if Mike finds him,' he replied, and he had scarcely spoken when the bird was flushed by the dog and flew over some brambles and stunted gorse, but not quickly enough for safety, for the big dog darted after it, leaped nimbly in the air, caught it deftly, and brought it to his master. The shepherd took it, but it died in his hand. 'Reckon you diddun want see *that*,' remarked my friend. 'I knows you, you see; I knows you likes t' see 'un *fly*, an' not see 'un *die* – though a landrail be good eatin'; – at least *I* like 'em ef they be cooked proper.'

When I left the shepherd that day I wished to visit somebody else, and asked him to suggest a way across the Downs. He showed me a little-used path. 'You goo on till you comes to a stile,' he said, 'then over it into a hollow – 'tis like a pudden-basin when you gets down in – an' then out t'other end over 'nother stile. These ol' paths be very handy,' he continued, 'but you want to know 'em! Now afore you goo I'll show you 'nother ol' path.' He led me to a big gate and pointed over a ploughed field. 'That *was* it,' he said, 'but 'twas ploughed up, an' now nobody knows no different. *I* shouldn' goo now, you unnerstan', being shepherd where I be, but ef I lef' here, an' comed back anywhen for a doddle round, an' cared t'goo across, there bean't anyone as could stop me!'

Such scraps of knowledge, hidden in the heads of shepherds and other workers in the country-side, are of increasing value in these days, when pedestrians have been practically driven off the main roads.

I followed the shepherd's directions and climbed the stile, and was soon in the 'pudden-basin.' I have been to lambing-folds tucked away in cosy hollows between hills, but this was different. I had no idea that such a place existed. Nothing was seen but turf and sky. The big, round, deep hollow was a place for day-dreams, and I lingered there for some time. It was a strange experience to lie there and watch some rooks pass overhead, for the depths of the green bowl gave them an unusual effect. At last I climbed the second stile and followed the path according to the shepherd's directions.

My habit of tramping about the country-side and using old footpaths, combined with my interest in flints and many Downland subjects, earned me a local reputation. I was apparently considered to be mixed up with all the societies interested in archaeology and downland and footpath preservation,[13] for while chatting with a man who was mending a sheep fence on a farm he mentioned a certain hill and said: 'Now, take that hill. We allus thought it *all* belonged to Mus Blank. I don't know why, I suppose 'cos he never said it didn't! – anyway, we never took t'sheep up there. Well, one day it come to it as we foun' out t' truth when *your* lot come 'long!' The word 'archaeological' was apparently too much for him, for he continued: 'They come doddlin' roun' t' hill, an' measurin' up, an' foun' out as t'top o' t'hill warn't hissen at all! Now, nobody wouldn' ever a-known it ef these 'ere Archipeligo blokes hadn't gone ferretin' roun' there.' He stopped to watch a bird until it disappeared, then he remarked: 'Yes, I reckon they be useful, knowledgeable blokes, an' I don' say nought agin 'em.'

My fame as an 'Archipeligo bloke' seemed to follow me round, and many shepherds put by various things for me. Bells and tackle, crooks and shears, 'shepherds' crowns,' flints and ox-shoes, thatching needles, lanterns, bows, and many other things were my reward for sincere interest in the men and their work.

Shepherds have very sharp eyes and find many odd things on the Downs. Among the treasures saved for me were four polished flint axes from different localities. The best one was found by Ted Nutley at Pangdean. Albe Gorringe, of Brown's Farm, West Blatchington, also surprised me with a 'shepherd's crown' which he termed 'a curiosity one' – a rare specimen, with the shell of the creature on it almost intact.

I have a store of such things, which a broker's man would despise, yet to me they are more than curios from Downland farms, more than souvenirs of wonderful hours, for they link me with men whose never-failing kindness was a revelation.

The Wind on the Downs

What a wonderful thing is the Downland wind!

When the warmth of a summer day becomes oppressive in the valleys it is only necessary to climb to the nearest brow to be refreshed by the little breezes always present there. Some of them come a long distance over the Downs and play first in the clover-fields and bean-fields. Others on their way to us dance over the scented turf, over sweet honeysuckle hanging on the thorns, over brambles and thyme carpets and a hundred other beautiful things, and from each one they bring particles of sweetness. So soft and gentle and caressing are they that we bless them for their welcome presence on the hills.

There are times when storm-clouds tempt us to wander out and enjoy the glorious colour-schemes in the sky and the soft, misty blue haze in the distant views. Then we often meet the south-west wind. A bluff, strong fellow, this, who blows upon us and chases us and tempts us to dance along, with arms outspread, over the Down. He would like to make us all into children, ready to hop-skip, hat in hand, while he runs after us. It matters not whether our age is five or fifty-five – the sou'-wester offers himself as a playmate, and if we are fortunate enough to be going in the right direction he will push us home ere the rain catches us.

Even on fine winter days when Downland ramblers enjoy new aspects of the paths trodden in summer-time the wind on the hills imparts a freshness which is exhilarating and it is only when it blows from the north-east that we learn the advantage of keeping to the sheltered valleys. Beware of the north-east wind! He is the bad boy of the family. He has a bad reputation, and deserves it. When he is about he seeks for those who boast that they love the wind and call it 'Heaven's music.' Hoping to catch these people unawares, he sweeps across every hill-top. You cannot avoid him; with his whip he slashes at every one as he rushes by. Sometimes people defy him. They button up their coats and jam down their hats. They walk against him and say they are not afraid; but should these intrepid folks persist in their defiance he invokes the aid of that giant of awful power – the Wind-God of the Downs. Then, on the hill-top, the giant lies in wait, ready for the braggart coming up the slope. The shepherd takes warning and retires with his flock into a sheltered bottom. He is a wise man.

This morning was cold, grey, and wintry, and a north-east wind was blowing, but I set out over the Downs, for I had an appointment with a shepherd. I defied the wind. I buttoned my coat and jammed down my hat and mounted the slope, for it was necessary to go over a ridge to get into the valley I had arranged to visit. Many a time I have wished for a physique equal to that of a shepherd, but even

Henry Rewell succeeded Tom Rusbridge as shepherd on Ditchling Beacon. His trousers are tied above the knees to keep them out of the mud. He had a keen sense of humour and once told a lady visitor to the farm that 'you tie them round like that to keep the rats from running up'. A frail can just be seen hanging on his back

with that it would have been impossible to do battle with the giant who waited my approach. As I reached the brow he caught me. He shook me till I was breathless and pushed me across the track. I grabbed my hat; I bent low and tried to dodge him. No use! Time after time he caught me and in his fury smote me and dashed me down! There I lay at last, exhausted and dazed. I crawled slowly over the turf, half blinded by tears and the force of his blows. My handkerchief was torn from me and flung a mile away over the valley. Very gradually I crawled down the slope and regained my footing; then I hurried to the bottom as fast as I was able. There, in a bend, I spied the shepherd and his flock in the most sheltered spot that could be found. Even he was in the lee of a stack for warmth and refuge while he waited for me. Fortunately the fold and sheds were near the end of the bottom, so we went there without delay and I was glad to accept the offer of a seat in the shed.

The shepherd's mate was busy in the shed fitting a stick to his crook – a burly man, quite civil, but a trifle morose and rather deaf. 'Joe,' bawled the shepherd, 'Mus. Wills was blowed down comin' o'er t' brow.' 'Right day for ut,' growled Joe. 'Us all ought to be big 's Joe,' remarked the shepherd. I said, 'Yes, indeed we ought, but I don't think Joe could stand on the brow to-day. I have had enough of it for once; yet there are some people who say they like a rough wind.' Then the shepherd bawled again: 'Joe, Mus. Wills says he knows folks as says they likes t' rough wind.' Joe looked at me and grunted; then he said the longest sentence I have ever heard him say: 'Hap they likes what *they* calls wind, but if onybody told me as they likes *this* ole wind I reckon I'd call 'un a liar!'

To-night I am sitting by the fire after my battle with the wind. Now I can hear him outside and once more I defy him. I can afford to laugh at his noise as I recline in comfort while the log crackles and splutters. What a joy is a good log fire! It turns my room into a magic cave. The subdued light from a candle shining through the horn windows of the old lanthorn is sufficient addition. The perforated lanthorn top throws a pattern of light on the ceiling. The furniture seems to recede into dark shadows, the sheep bells are only masses of shadow, the shepherd's crook and other things are scarcely discernible. Then as I raise the log with a poker the flames leap up. The inside of the ancient copper bowl with a hundred dents instantly becomes a glowing furnace of red-gold! Tiny flame-like points of light flicker over pewter and brass. In the bowls of the brass ladles they wink and twinkle as do the stars. The light jigs and skips up and down the highly polished barrel of the crook and dances merrily over its curly guide. High lights on the glaze of the Sussex-ware[14] jugs are turned to molten drops that fail to fall. Glass and china, wood and metal all seem to be alive; the whole room has warmth and colour and life when I have the company of a good old-fashioned fire. The window is shut, the door is barred, and in my chair I am safe from the fury of the giant who followed me home!

— THE SHEPHERDS OF THE — DOWNS

The Lure of the Shepherd's Life

When the blackberry harvest draws people from the towns the shepherd's territory is often invaded by folk who go to pick in the same spot each season. It is their annual excuse for a 'day out,' their one opportunity to experience that sense of freedom which only the country-side can offer. Unconsciously they note the presence of rooks and other birds, of butterflies, of cattle, or a flock of sheep. A casual glance may reveal the shepherd, with his dog in attendance.

'There's that old shepherd again!' I heard one woman say to another, 'still standing about! He looks just the same as when we came last year!' So he may do! So do the rooks, the butterflies, and the cattle; so do the brambles and the familiar view around, and though the berry-pickers are a year older, and must have done a certain amount of work during the year, they are apt to forget that the shepherd has any hard work at all. He may look 'just the same' as when they came last year, but few realize what he has done in the interval. Lambing time, when he was on duty day and night, has passed again, with all its usual round of labour and anxiety. Other jobs have followed: tailing, cutting, shearing, dipping, and trimming. Day by day, in all sorts of weather, he has tended his flock. (What a multitude of duties are included in the word 'tended' – you cannot realize it unless you have been in his company constantly.) Now he stands, apparently idle, but it is safe to assert that his thoughts are centred on the ewes, for on his care of them during the winter months depends the success of next lambing time.

So each year passes, and the shepherd carries on, working seven days a week. In return for such devotion he receives a wage which other workers would despise. I have known a farmer to say: 'I wouldn't part with old Mike; no matter when I go by, early or late, week-day or Sunday, he is always about, looking after the sheep!' Perhaps he imagined that old Mike worked simply to please him, but it was not so. Such work is not all done for the sake of the farmer, for if the shepherd moved to another farm he would do just the same. It is the lure of his craft which causes him to put his dumb families before everything else, and to stay with them for many an extra hour in order to give a 'last look round,' for their benefit.

Many references have been made in print to the lives of shepherds, and many opinions expressed concerning their outlook on life. If we were able to interview all the old Sussex shepherds at this date we should probably find that the majority of them view the changes and ways of modern times with a certain amount of disfavour. They are forced to adapt themselves to new conditions and move with the times to some extent, but the habits of a lifetime (in many cases strengthened by hereditary tendencies) are not easily altered.

My hours spent with shepherds at all seasons of the year have been a revelation in many ways. My own early idea of a shepherd, as a man, crook in hand, with nothing to do but to watch his flock, was soon exploded. Gradually I sorted out the many little tangled lines of thought that occurred to me; gradually I traced the reason for some action or for the careful attention given to some little detail, and every fact I gleaned pointed backwards – back through the years to the days when the shepherd learned his craft, and sometimes further still, to the time of his parents and grandparents. One old man explained a point thus: 'My gran'father did it thet way, an' my father did it thet way, an' now *I* does it the same. There be a right way an' a wrong way o' doin' things, an' the sooner you learns to do a thing the right way the better it be for ye, for then yew see yew doos it right wi'out thinkin' 'bout it.'

Many of the oldest shepherds now living started work at a very early age. They absorbed the details of their craft at a date when the craze for speed was unknown; when, as one old man remarked to me, 'the farmer shared the shepherd's pride and delight in the flock, and in all the shepherd did.' That is why we still find old men doing more than a young man would expect to do for the wages paid. With experience of duties gained as shepherd boy, teg boy, and under-shepherd grew that pride which was shared by the farmer – a pride which even extended to the bells used on the flock.

How different it is today![1] The craze for speed has affected everybody. Ways have changed; good manners are old-fashioned; thoroughness is out of date; the old shepherd himself is out of date; few care twopence about his thoroughness or his pride in work; his only enjoyment is in the thought of past days when he was 'somebody,' and so any old shepherd you meet is usually quiet and reticent until he discovers whether your interest is genuine. If he is satisfied you soon know; he responds to enquiries, but he is like a dormouse waking up in spring, for at the touch of memory's hand his attitude changes, his stiffness relaxes, his eyes are young again, and as he talks for a little while he becomes 'somebody' once more!

Do not despise an old shepherd because he appears to be simple; his simplicity is actually a sign of strength. He may not fit in with modern ideas, but it must be remembered that when he started with an inborn love for one kind of work he was first taught obedience, correctness, thorough-ness, self-reliance, and untiring devotion to the flock. As shepherd he became a man of importance and ruled his own little kingdom. The drawbacks attending his job were the long hours, the many inconveniences resulting from unkind weather, and the low wages; but there were compensations in other ways. His rather hard upbringing helped him to tackle every task in a methodical manner, without unneces-sary worry, and his solitary peaceful hours in the open air gradually cast a spell over him – a spell so strong that at last he developed this quiet, simple manner. Hidden under this

Another year begins for the shepherd and his flock, secure in the straw lambing fold

'As lambing-time approaches, the site for the fold is chosen. Sometimes it is built near a rick, so that the necessary supply of hay or litter can be procured easily. The coops, placed in rows, hold the sweet-scented hay, and the sheep stand all along, thrusting their heads between the wooden bars to enjoy the meal provided. The shepherd's hut is placed in position near, for he must be on duty day and night, with short intervals, from the time the first lamb arrives. With his hand lamp he makes his round during the dark hours, giving any assistance he can.'

cloak is a wonderful love for the freedom of open spaces, for pictures of sheep on the hillside and in the fold, for fields of mangold and rape, for the familiar starlings and the hundred and one items which he meets and notes as he goes to and fro.

Dog, crook, and favourite bells are still his principal possessions. His pleasure is in his flock, and, happy in the many memories of past days, he jogs on, impatient with new farmers, with alterations in farming methods, and with golfers and motorists who invade his once quiet grazing grounds. Gone are the old sheep-shearing suppers in which he took part, – gone are the old songs in which he joined, – gone is the farmer who was human enough to compliment him on the chimes from his old sheep-bells. Gone, too, are many other shepherds who were his friends when he was young. They have passed away, and he admits that he is one of a dying race, but in spite of all this he carries on his usual daily work with the same unwavering persistence – especially if he be a man from a family of hill-shepherds.

The term 'shepherd' embraces those who tend flocks on the hills, on flat farms, and on the marshes. At one time I thought that the hill-shepherds considered themselves a peg above the others, because some had owned that they would never leave the hills to look after a 'farm flock,' but when I had been in their company a hundred times, at every season of the year, I understood why the lives of these old-fashioned men are filled with romance, although they do not appear to be aware of it, and why a 'farm flock' would not satisfy them.

Stand with an old shepherd on a hill-top on a fine summer day, and you will capture some of the beauty which has moulded his character and that of his ancestors. He is king of a beautiful and peaceful little kingdom. The purest breezes caress him. The music of his precious sheep-bells delights him. As he looks across to other hills and into the valleys and watches the sun chase the cloud-shadows over the turf he may tell you, in a simple manner, that in the same way a fine day on the hill-top chases away remembrances of hours of work in bad weather. He may tell you of bygone days when he caught glimpses of other shepherd friends on distant hills and how the wind brought the sound of other bells to him. The flower-studded turf and the wild life around are more to him than any book could be. You may count the varieties of blossoms and name them all, but names out of books mean very little to the shepherd. All the flowers are friends that he has known from his earliest years. There are moments when you feel that his love for the hill-tops is beyond description in words, and, in a vague way, that such a man is superior to the ordinary mortal whose life is full of trivial affairs. He has partaken of the peace of height and expanse, the beauty of sunrise and sunset, the company of birds and flowers and grass, and generations of sheep for so long that his whole outlook on life is far above your own. The poor old hill-shepherd may have given lifelong work for a meagre wage, but he is rich in many things that money cannot buy!

Some Downland Shepherds

Although a book could be filled with portraits and biographies of shepherds the accounts of the few recorded here must represent the whole number. These have been specially selected to show that actual interviews have resulted in the gathering of facts relating to old days and ways and valuable notes on various subjects. My pile of note-books proves how numberless visits to shepherds resulted in the collection of a store of information, without which it would have been impossible to compile such a book[2] as this or to obtain a true and intimate knowledge of their lives.

NELSON COPPARD

Nelson Coppard was born at Poynings in 1863. His father was shepherd on Dyke Hill for eighteen years. He started work very early as shepherd boy at Horton, near Beeding. Then he became teg boy to Eli Page, of Patcham, and later he served as under-shepherd at Saddlescombe. Since then he has been shepherd at Trueleigh, Iford, America (between Firle and Newhaven), Balmer near Falmer, and Mary Farm, Falmer. He is now at Pangdean Farm by Clayton Mills.

To write a full account of Mr. Coppard as I know him would be a long story. He was the first shepherd I ever met. From him I had my first instruction on sheep-bells, crooks, and the details of a shepherd's life. To the lucky natives of Sussex a meeting with a shepherd is just an ordinary incident in a Downland ramble, but to a Londoner, blessed with an artistic temperament, that first sudden entry into a little valley full of sheep with their ancient bells chiming, the meeting with a jovial shepherd with his glittering crook, the chat with him, and the return journey, when I carried home wild flowers and two large canister bells, was over-whelming. I felt that I had stepped into a new world.

My anxiety to learn amused the shepherd. It was a fresh experience for him to find anybody so eager for the informa-tion he could impart. Though I have since found many shepherds and gathered a store of memories, I do not forget that our first meeting was the start of a long series of rambles which have at last ended in the making of this record.

There is always something of interest or something quaint to note down after a visit to Nelson. To-day, as I stood on the hill by Pyecombe Church, I could see his fold at the edge of Pangdean Farm. I was not expecting to hear much music from sheep-bells, knowing that it is his custom to remove most of them and keep them in his hut at this time of year; I was therefore startled by the deafening burst of sound which greeted me as I approached the fold. The din did not stop until I actually passed the first barrier of wattles and furze branches. Behind these I found the shepherd and his son-in-law, Ted Nutley, holding the bells and laughing heartily at their success in 'ringing me in.' 'I was just taking the bells into the hut,' explained Nelson, 'when Nutley saw you coming along. I know you like to hear the bells when you come, so we rung 'em for ye!'

After dinner we inspected the flock. The shepherd pointed out a dead lamb. It was well grown, but had collapsed and died suddenly, as is sometimes the case. We fetched the ewes from another fold. The flock spread out, as mothers and children, making a babel of cries in various keys, sorted themselves and were gradually united. One ewe, wandering about and calling plaintively for her baby, at last found the body of the lamb. Nelson approached, crook in hand; caught the ewe and examined her. 'Thought so!' he exclaimed; 'she be full of milk!' and he led her into the fold and secured her in a pen. Then he sharpened his knife, fetched the dead lamb, cut through its skin round the joints of the forelegs and slit down the chest, and in a few minutes off came its woolly jacket, quite clean and inside

'The Changeling'. Habberton Lulham's picture shows an orphan lamb being fitted with the skin of a dead lamb by Henry Rewell so that it can be fostered by the dead lamb's mother. 'He be like we when we gets a new suit – it don't feel jus' right at first.'

out. This was at once turned, fitted with strings and hung on a stick ready for use. A lamb that was not getting milk owing to the temporary sickness of its mother was caught and dressed in the dead lamb's skin. He was quite big enough for it – it fitted him as closely as a suit of combinations when it was tied on. The lamb's movements provoked a smile, and Nelson said: 'He be like we when *we* gets a new suit – it don't feel jus' right at first!'

The ears on the skin gave the baby a quaint appearance; at first sight one might have thought that a lamb with four ears was on show. Now began the troublesome task of persuading the baby to feed from its foster-mother. After many patient efforts the shepherd succeeded, and although the ewe sniffed suspiciously at the lamb at first the two had accepted each other before we left them.

Mr. Coppard's flocks have varied greatly in number on various farms, but he states that in a general way the flocks kept now are much smaller than in past days. Many other changes have taken place. He has no use now for the big dipping hook which he used at the sheep-wash, as this part of the shepherds' work has died out in most districts.

I questioned him about sheep marking. Ordinary marking on the wool with a stick dipped in colour has been his usual practice, but he has had many sheep ear-marked with holes and snicks in the ears, also some few pedigree sheep which had 'ear-rings' – small brass or other metal tickets fastened to the ear by a ring. Tattooed marks inside the ear stamped with a special punch are now found in Southdown sheep from registered flocks. A ewe was caught for inspection. The flock number, '550,' was stamped inside, and there was also a round hole punched through near the tip and a triangular snick taken out of the edge. Punched marks are often used as a record of age, and are a more reliable guide than the animals' teeth.

I was collecting notes about shepherds' clothes and Mr. Coppard stated that good corduroy suits and gaiters, with a hard felt 'bowler' hat had been his usual dress, but his father wore a smock, 'a blue one,' he said, 'not like the slaty-colour one I got for you, but blue – what you might call a *butcher blue*. He always wore that over his corduroys, and in bad weather an overcoat on top o' that! I only wish I had one o' those overcoats,' he remarked, 'but you couldn' get such a thing nowadays. They were thick and rough and fleecy. I remember my father wore a *white* one, though there were all sorts about. In my young days a shepherd could sometimes get hold of an old cavalry cloak. They were fine things to keep ye dry!'

Nelson is noted among other shepherds for his fondness for sheep-bells and good dogs. He says there used to be far more dogs of the rough-haired type in his young days, and fewer collies than at present. A stranger once said to him: 'Your dog obeys you well, shepherd – you must have payed him pretty much to get him to obey you like he does!' Nelson could hardly believe that he had heard aright. At last he said: 'If I "payed" *you*, as you calls it, would you do any work for me afterwards?' 'No, I wouldn't,' said the stranger, 'I guess I'd keep out of your way.' 'Very well, then,' exclaimed Nelson, ''tis the same with a dog – you got to teach by *talking* to him, *not* by *paying* him. You'll *never* teach a dumb animal to like you by *paying* him!'

The shepherd has a keen sense of humour, and a very dry way, as those who chat to him soon discover. He told me of his interview with 'an old grey-whiskered gentleman in riding kit' who had just returned from a hunt. 'Dear me, shepherd,' he said, 'my feet ache and my legs ache so much I don't know how to walk.' 'Well, I suppose you've bin on 'em a good time, haven't ye?' asked Nelson. 'No,' said the hunter, 'it isn't that, for I've been riding all day!' Where-upon Nelson remarked, in his usual dry way, 'You don't get my meaning – I mean you've *had 'em* a good long time, surely!' – and then the old gentleman saw the joke.

Other little tales and comments passed the time away. Nelson had saved some old ox shoes he had found and told me that black runts,[3] used for ploughing and other work, were once bought and sold at Steyning. Then the mention of Michael Blann's name recalled the fact that Mr. Blann once cut out a sundial in the turf for him to use. He was shepherd boy in those days. He stayed out all day and was provided with his lunch. This was generally the top of a loaf. Sometimes it was pulled open on the soft side and some butter put in it, but if there was no butter he had a piece of fat bacon out of the brine crock. He recalled a certain day when he sat on the brow of a high hill to have his lunch with a companion. The loaf-top slipped from his hand, and rolled and bumped and danced all down the slope 'like a cannon-ball'. They both laughed to see it go. It was too far to fetch it, and Nelson lost his lunch, but the loss was forgotten in amusement at the incident, and to-day the thought of that rolling loaf-top still brings a smile to his face.

It is refreshing to meet with anybody so outspoken as this shepherd. 'I think it be the best way,' he once remarked to me. 'I says what I thinks, an' I talks in front o' people as I talks behind their backs! If what you *thinks* be right then, what you *says* will be right too!'

JACK COX

It was a fortunate day for me when I tramped over the hills from Washington to Lee Farm and found Jack Cox, the shepherd. Such courtesy! Such readiness to listen to my request for information! Such willingness to assist!

Jack Cox, shepherd of Lee Farm, near Harrow Hill

On the green bank opposite his house I explained my errand, watched all the while by Jim, a beautiful white sheep-dog. Our conversation soon drifted to details of shepherd-craft. Mr. Cox has a Pyecombe crook, light in weight, and about thirty years old, so this must have been made by Mr. Mitchell. It is rather a poor specimen now, but it is still the shepherd's favourite. His son had a nice new crook of local make, but Mr. Cox does not covet it. He tried it for the first time in front of the farmer (a practical man) and missed a catch, much to the farmer's surprise. He explained about the new crook, and returned to his old 'Pyecombe.' He said: 'Directly you have tried a crook you know, somehow, if it will do for you. Now, this old hook will hold any sheep for me, and it has never failed!'

The shepherd used to wear a smock, and told me a fresh item of interest in connection with shepherds' dress. He has known shepherds, when beginning work, such as fold pitching, to change from smock and hard hat into a rough sacking slop and a red cap ('well, a little cap, anyway, most times a red one'). 'A hard hat was in the way when carrying hurdles, which would sometimes touch the hat and tilt it over your eyes.'

I was anxious to see the shepherd's old canister bells, for, as I told him, I had listened to them on a former visit to Harrow Hill when I was hunting for flint implements. My joy in the sudden discovery of a beautiful flint axe while the songs of the bells were borne to me on the breeze made that ramble a memorable one. To-day the bells were silent; they had been removed for shearing time, but the shepherd took me to the lambing yard to see them. In the corner was a little room, with hearth and chimney. His arm-chair was there – a low-backed one. 'This is how I sit before the fire at lambing time,' he said. 'My father taught me to do it,' and as he spoke he demonstrated for my benefit. 'If I

doze,' he said, 'my head wags and drops, and I start up and know 'tis time to move again, or I might sleep too long. My father said that was the way. 'Sit orkard,' he used to say, 'sit orkard an' you won't sleep too long,' and so I always do it!'

The precious bells were hanging in the room, and while I inspected them he told me that in war time he had a flock of nine hundred to look after. His wife was his only assistant, so she donned shepherd's clothes, and between them they did all that was required.

The shepherd enjoys his life among the hills. 'I wouldn't live in the town,' he said. 'If you want a pound of potatoes you go to a shop and buy them; if you want a few sticks of wood to light a fire you go to a shop and buy them! No, the wages may be less in the country, but you save in many small ways, and you live in the finest air there is!'

Before I left, he allowed me to select a bell for my collection, and I chose an old well-worn canister with a wooden yoke. 'These bells,' he said, 'came from a shepherd named Michael Blann, years ago. *He* was one of the old shepherds, and ought to be included in your book. If he is still alive he must be old; he was living at Patching when I last heard of him.'

I packed the precious bell in my satchel – a souvenir of a very pleasant visit to Lee Farm – and wandered home over the hills.

At a later date I met the shepherd again as he was returning to the farm with his flock. The dog Jim was with him. 'Don't touch him,' said Mr. Cox; 'he is rather sore about the head.' They had found a badger, abroad in daylight, a few days before, and when it ran into a little hurdle pen in the corner of the field and tried to hide in a hole in the bank Jim attacked it. He received a nasty bite, but the badger was killed. Mr. Cox remarked that it showed he must be a strong dog and very plucky to have tackled a badger so fearlessly.

As we stood listening to the songs of the canister bells on the sheep he talked about the scarcity of shepherd boys. He said that at present there are plenty of shepherds, but so few boys training for the work. The mere minding of sheep for a few hours is simple enough, but there is so much to learn before one can be called a shepherd. He mentioned a man who thought the shepherd's life an ideal one, and willingly tried it for a day. He accompanied Mr. Cox early in the morning. He found it difficult to pick up three hurdles on a pole, to carry them and put them down correctly. 'They went round on the pole like windmill sweeps,' said the shepherd. His next experience – that of pitching a fold – was equally disappointing. The iron bar seemed to make so many holes in the ground, instead of only one big enough for the stake. 'That was only *two* things out of the day's work, as I told him,' said Mr. Cox. 'All that I told him to do *I* learned forty years ago, and every year since I have learned something fresh about sheep that I didn't know before! That is why a man of my age could not start as a shepherd now – he would always be forty years behind!'

MICHAEL BLANN[4]

I asked a little girl in a Patching lane if she knew where Mr. Blann lived. She answered: 'Yes, sir, he lives in the first cottage on that side.' Then she added, very seriously: 'He's got a long beard, right down to here,' and she touched her little knees. Thus when I arrived at the cottage and saw a man just coming to the door from the meadow at the side I knew that I had found the shepherd I sought, although the length of his beard was less than I expected from the child's description.

We were soon chatting, and I learned that Michael Blann comes from a Sussex family. His father was born at Sompting, and Michael was born at Beeding in 1843. Although his family were not shepherds, he was put to the work when nine years old, and stayed at it. His wages were 3s. 6d. a week, although at that date some boys received only 2s. 6d. A shepherd's wage in his young days was 12s. or 13s. a week – 'not much when you come to think of it,' he said. True, he had his house as well, sometimes, but even then to bring up a family on such a wage wanted some doing, and there was not much left for the shepherd!

We spoke of Findon Fair. 'It is only a sheep sale now,' said Mr. Blann. 'I remember the first time I went to Findon Fair – seventy-seven years ago, when I was only ten years old. I had to mind the flock up at the end of the field while shepherd went to find the pens he was to use. Near to me was a thing like a round table with hooks all round it. It was set near to a big fire. A pig was cut up and the joints were hung on the hooks. The big turnspit took the joints round and round, and when they were cooked folks bought roast pork for their dinner.'

While speaking of clothes worn by shepherds I told him of Mr. Cox's reference to caps worn when pitching, and he confirmed this, as he wore one for the same purpose. Referring to wattles hitting a shepherd's hat, he said: 'Some shepherds had dog's-hair hats – you don't see them now! They were something like a billycock, but very thick and very strong. They were heavy to wear when working or in the sun, and so a cap was carried in the pocket.' One short man he knew who had occasion to reach something high up stood on his hat to do it!

The shepherd wore a round frock for many years, and

Michael Blann, best known of all the Downland shepherds for his singing and love of music. This photograph was probably taken at Patching, towards the end of his active shepherding days

longer than some who felt 'old-fashioned' in one. He told me of the large, heavy overcoats. They were fleecy outside, and very thick, with a big cape. He was once fortunate enough to own a white one. When it was at last worse for wear he turned it very carefully and sewed it together again with twine to make it last as long as possible. His wife made his smocks for him. The last one was made to his own idea to open all down the front. 'I somehow got tired of pulling it off over my head,' he said, 'although everybody knows that is the way to get out of a smock.'

Outside the cottage lay the remains of a big cask. It must have been old and dry, for when a passing waggon touched it it collapsed. 'It all fell to bits in once!' said Mr. Blann. (The thought of it always amused him.) He moved the pieces with his stick, and revealed several short bits of curved branches which he had cut for making sheep-bell yokes. He showed me how some could be split to make two yokes instead of one. (How deeply must the charm of bells bore into a shepherd's heart if it causes a man of eighty-seven to cut yokes which he will never use!) 'I was very fond of my bells,' he said, 'and very particular to keep my tackle in good order. I had one little canister which I once heard from two miles away. I was always fond of it after that. I cut out all my own wooden crooks (bell yokes) and bone lockyers, and took great pride in them. Now Jack Cox has them all; he has had them a good many years, so he knows they were good bells and tackle.'

In reply to my enquiry, Mr. Blann said he had never seen bells for sale at Lewes Fair. 'I reckon that must have been a good time ago,' he remarked.

I also asked about his crook, but learned that he had parted with it long ago to 'a man who took it away to London.'

According to the shepherd's opinion, modern 'progress' has not improved us. His comment was, 'People are not so content nowadays, even with all they've got! They're always rooshing about, and yet they have time to pick one another to pieces! They don't try to help one another, or put one another together a bit!'

Mr. Blann loves music. At one time he could play several instruments. He was well known as a singer, and often sang in company with a friend. 'I enjoyed that,' he said, 'for we got on very well together.' He would sing alone if asked, and was often called upon for a song at fair 'sing-songs,' shearing parties and local entertainments. He has a book in which he wrote the words of all his favourite songs. It was started in 1867 and is a precious possession.[5] It was once borrowed by a collector of old songs, who would not return it, but another enthusiast turned up, and hearing that the first man had kept the book for years, went after it, and at last the shepherd had his treasured book again.

Mr. Blann also has his favourite flute, but finds it less easy to play than formerly. He was surprised at my delight when he produced another treasure – a tin whistle which he always carried in his pocket, so that he could play to himself on the hills while minding a flock. It was also useful at times for pitching on the correct note quickly when starting a song. I sat by him while he sang a verse of four of his favourites to me and played the tunes slowly on the whistle-pipe for my benefit. I listened to 'Sheep Shearing,' 'The Harvest Home,' 'A Drop of Good Beer,' and 'Rock the Cradle, John!' and thus I was linked for the first time with the old sheep-shearing suppers of years ago when the gang called on Michael for a song.

I had read of shepherds using reed pipes and whistle-pipes, and had found pictures portraying such scenes, but it was delightful to one engaged on a Sussex shepherds' book to prove an actual instance in the country.

it and look out at the flock. It would have taken three days' rain to wet it through.'

At the mention of a smock as a protection in bad weather he told me some interesting facts about the shepherds' dress in his boyhood. 'When I was a boy,' he said, 'I had three brothers. Father and us four boys made five to dress, and we all wore corduroy – buff colour – and corduroy gaiters too. We all wore smock frocks over that – blue ones. My mother made all the clothes (and looked a long time at a shilling before spending it), but the stuff was good then! We couldn't afford leather gaiters; she made us cord gaiters to match the suits and lined them with canvas, and they kept out as much wet as the leather gaiters I buy now do! Every year, after harvest, mother went off to the shops and bought a big roll of corduroy, and canvas, buttons, thread, and things, and we all had a suit in turn. She made our smocks, too, all tuckered up proper, with big turn-down collars, and pockets that 'd take a rabbit easily without showing it if you walked upright. Overcoats? Well, yes, we did have one of a sort, but anything had to do. We had good suits and smocks, and there wasn't much money left for overcoats.'

CHARLES TRIGWELL

At Findon Fair I met Jesse Moulding of Goring, the last of several generations of Sussex shepherds. He sat on a wattle by his sheep, for the sale had commenced. 'Do you know Charlie Trigwell is here?' he asked, and pointed to where this old shepherd rested while he surveyed the scene. I took the opportunity to act as messenger, and in a few minutes Mr. Trigwell came back with me for a chat. He used to be a well-known figure at the fairs until ill-health put an end to his work, and this time he was only present as a visitor.

Shepherds Charles Trigwell from Shoreham, born in 1851 and 'noted for his broad smile', and Jesse Moulding from Goring, the last of several generations of Sussex shepherds

I told him about my shepherds' book, and before I left him arranged to visit him at Shoreham.

The shepherd has a treasured sheep-crook made of brass,[10] which he showed me when I went to Shoreham. It belonged to his uncle, Tom Trigwell, and was the first, or one of the first, made, as his uncle's crook was borrowed by the maker of the brass crook for a pattern. I had heard of these brass crooks several times, and only three weeks before I had found that Mr. Wooler of Pyecombe knew the maker. I experienced a thrill of satisfaction as I handled this specimen. It is well made and nicely finished and must have been a pretty thing in use in the fold.

Mr. Trigwell was born at Hove in 1851. He is noted for his broad smile. His keen sense of humour makes him an entertaining companion, and I was kept amused by his reminiscences of people and quaint incidents. When I asked about smocks he said: 'Yes, I always wore one, and a billy-cock hat like I wear now. My uncle Tom wore a smock too. He was a rare man for "rape greens," and when he picked a mess of greens to take home (the usual shepherd's allowance) he used to tumble them into his round frock as he picked them.' The shepherd's smile broadened at the thought of him. 'Him and his greens!' he exclaimed. 'Once I was ready to go home – and glad to go too, for it was miserable old weather. 'Come on, Uncle,' I said, 'I'm ready!' 'Stop minute!' he says, 'I ent got my greens yet!' Diddun matter how miserable a day 'twas he wouldn't go without his greens!'

The mention of rape greens reminded me of Findon Fair, where another shepherd had praised a big variety of kale with thick stems as sheep food. Mr. Trigwell agreed, in his own quaint way. 'Of course they like good food,' he said, 'they're only like we! We don't want a bit of apple peel or orange peel – we like the fruit!'

He told me two little tales of his Uncle Tom, who was a well-known shepherd fifty years ago. One of them also includes Shepherd Clem, famous among all his friends at that date for his extraordinary appetite. Tom Trigwell and Clem once dined, at fair time, at an inn where an excellent half-crown dinner was provided. As they left they saw a woman selling penny pies, and more in joke than in earnest Tom said to Clem: 'I'll pay for all the pies you can eat after all that dinner, Clem!' At once Clem started, and ate thirty pies, a feat which emptied the basket. Tom had to pay the half-crown, and was rather vexed. When the woman offered to go and get another stock of pies from home he refused to wait, and cried: 'Good Lord, no! – doänt get him any more, missus! – he'd eat more, an' then eat the basket, but I reckon he's had enough!'

The other tale was an account of how three 'roughs' were paid in their own coin. In explanation Mr. Trigwell said that the Downs were not as safe at one time as they are now. There were more 'rough characters' about, who would choose quiet spots to pay off old scores or 'play pranks at someone else's expense, just for devilment.'

Tom's son happened to stop at an inn on the Lewes road, Brighton, one Saturday night, and while in the bar he overheard three young men discussing their programme for Sunday. 'Let's go up Moulscombe Barn, an' give t' ol' shepherd a roustin',' said one of them, and the others agreed. Little did they suspect that their plans would be known!

Tom was surprised when his son said: 'Wake me to-morrow morning, I'm coming up to the barn,' but he did so, and while on the way learned of the plot and planned a surprise for the visitors.

In due course the three men appeared and entered the fold. It was just after lambing time, and the shepherd had plenty to do to fill the cages with hay. The dog was tied up, and near to him was a pile of straw, under which Tom's son was hidden. 'The men messed about, looking at things,' said Mr. Trigwell, 'and Tom went on hayin', an' went on hayin', an' went on hayin' till they started their mischief. Then he told them to go.' This was the moment they had been waiting for, and what the end would have been nobody knows, but Tom shut the door of the fold and stood at bay with his back to it, holding his hay prong. 'Very well,' he called, 'if you wunt go out you shall stop in.' At this moment his son emerged from the straw. 'Come to give the old shepherd a rousting, have you?' he remarked, and he untied the dog. The men, taken by surprise, retreated to one side of the yard, but when the dog heard the order 'turn 'em back,' he obeyed instantly and they were forced to run. The shepherd repeated the order 'turn 'em back' till the dog had worried them well and nipped them by the breeches; then he called off the dog and opened the door. He did not have to speak again. The men raced to the door and darted through and disappeared, and the shepherd was not bothered any more.

Mr. Trigwell told me of the old days at the fairs, when flocks were taken thither a day in advance. Shepherds met at an inn at night, a room being reserved for them. On these occasions there was plenty of merriment, and 'colt-shoeing' was performed. 'Colts' were young men whose experience of shearing was limited to one or two seasons. The man appointed caught each one as he entered, lifted one of his feet and pretended to shoe him. The 'colt' was then expected to stand treat to the rest by buying half a gallon of beer.

With jokes and tales and reminiscences the time passed quickly. Mrs. Trigwell called me to take a cup of tea, and I was shown a beautiful photograph of Mr. Trigwell taken at Shoreham. My adjective brought comment from the shepherd. He remarked drily: 'Well! 'tis the first time anybody has called *me* beautiful!'

Before I left, the shepherd promised to leave the brass crook to me in his will. We parted with another merry jest, and the first man's face that I met in the street seemed to me a very miserable one after my interview with the old shepherd with his broad smile.

FRANK UPTON

On the farm known for many years as 'Brown's Farm,' at West Blatchington I found Mr. Frank Upton, a well-known Sussex shepherd. He is one of the old school. His father and grandfather were both born at Southernham, near Lewes, and both were shepherds, consequently his own training was thorough. Like most old shepherds he dislikes modern ways, and as I chatted with him our conversation naturally drifted to a comparison of the present day with the happier days of the past.

'You can't help noticing the difference wherever you go,' he said. 'Why, I remember when I could stand on the highest mountain in this part and see twenty flocks of sheep! There was plenty of work for the shearing gangs in those days,' he said, 'and plenty of good fleeces piled in that barn. The wool buyer used to come and offer so much a pound for them. Now prices are much higher. Old Mr. Brown once said to me: 'If I can be sure of thirty shillings each for my ewes I know I am making them pay me.' Now farmers say that sheep don't pay, yet they get bigger prices for sheep and lambs at the sales, and bigger prices for wool, and sheep don't eat more now than they did then, and their wool grows by itself the same as our hair do! The shepherd's wages

Frank Upton of West Blatchington: 'I remember when I could stand on the highest mountain in this part and see twenty flocks of sheep! There was plenty of work for the shearing gangs in those days, and plenty of good fleeces piled in the barn.'

don't amount to much – they never did! – so there must be something wrong somewhere! It is the ways of people now that is wrong. Farming wants method as well as work!'

Mr. Upton has a good memory for details of West Blatchington farm in the old days under John Brown, for he has been shepherd there since 1883 – fifty years. 'We had no cows then,' he said. 'Mr. Brown only kept one for milk – he didn't believe in mixing cows and sheep. We used to have about 750 breeding ewes and 300 tegs here. His other farms were at Patcham, Standean, Ditchling, and Rottingdean, and between them all he had always about 5000 breeding ewes. In those days the shepherd was in charge – why! if a shepherd couldn't take full charge of the flocks he wasn't a shepherd at all! John Brown would say: "Well, Upton, what about this, or that?" and it was soon settled! *Now* I mind these few tegs! I am just about to fold them – fold them on this stuff, sanfoin, not fit to fold on yet! How *can* they get a proper feed?'

The next minute he said: 'Oh, well, I've had *my* time and I've known the good old days and remember them! Our work was well done, and we just enjoyed Fair times! We used to have three days for Lindfield Fair – one to take the sheep and leave them; the next day we arranged the pens and so on, and afterwards we had a cricket match at Slug's Wash. Next day was Fair Day, and we came home after the sale.

'Fancy farmers giving shepherds a day off for cricket nowadays!' he exclaimed. 'Why, we only have sheep sales instead of fairs – there and back in a day, though *I* don't do it now – Albe does that – for I was seventy yesterday.

'Lindfield Fair was on the 5th August at one time,' he said, 'but as it sometimes happened to be Bank Holiday and it wasn't easy for people to get there it was changed to the 8th. Lewes Fair,' he added, 'was always on the 21st

September, and all us shepherds used to buy our clothes that day and we used to meet in the evening at 'The Swan,' Southover. If you could have seen inside there you would have found plenty of shepherds!'

Mr. Upton pointed to high ground near the Dyke road[11] and said: 'Years ago you could stand up there and look all round these slopes and see teams of oxen working. It was reg'lar picturesome! – it was the real old Sussex! Now, to us older men, 'tisn't like Sussex at all!' He then pointed to a long shed. 'At one time there were twelve oxen kept in that stable,' he remarked. 'Every year two were sold off and two new ones brought in. Now, take hurdles and wattles,' he continued. 'Every year we had enough new wattles for a fold. That kept the supply going, you see. They were good oak wattles[12] then, with iron bolts and rings, and lasted well. When I found one defective in any way I was supposed to put it by, and it was taken to the farm workshop and repaired. Iron tops and bolts were made good, and so we had no trouble when pitching. Method, you see! method in everything!'

In his quaint way the shepherd supplied the explanation of the present situation. In former days shepherds were consulted. Now they are ordered. Whereas they were once expected to say what they would require for the sheep they are now expected to bring flocks into condition on what is provided, which is not always an easy matter. Formerly only sheep suitable for the locality were purchased; now any sheep is expected to thrive anywhere and bring a profit in due course.

I have a treasured possession – a genuine old Pyecombe crook, made by Berry, which was given to me by Mr. Upton some years ago. Now, owing to his further kindness, I share his chief treasure and family heirloom – the charming picture of an old Sussex shepherd of a hundred years ago.

'A hundred years ago'. Frank Upton's grandfather, a shepherd from Southernham, near Lewes. 'Under strict conditions he [Frank Upton] lent me a wonderful portrait of his grandfather, taken at Hailsham. It is valuable as a record of the costume of the period.' Probably one of the earliest photographs taken of a shepherd, perhaps in the 1860s

WILLIAM DULY

Half of the September day had passed when I arrived at East Dean to seek for William Duly. I followed directions given me, but when I reached the top of a hill and scanned the valley and the opposite slope there was no sign of sheep or shepherd. Fortunately I met a lady who knew him, and she showed me where his fold was hidden away on a slope beyond the side of a cornfield through which I had passed. Satisfied to know that my forty-mile journey had not been in vain I followed the track to the fold, and found the shepherd just finishing his dinner.

Mr. Duly was born at Alciston in 1858. He comes from an old Sussex family. His grandfather and his great-grandfather were both shepherds, and several of the family followed the same employment; consequently his ways are old ways and he is convinced that his type must inevitably die out in due course. I suggested that the large fold he had to pitch was quite enough for him, but he made light of it. 'Flocks are not what they were,' he said, 'and won't be again. Farmers are putting down too much ground – at least a lot of it isn't put down,' he added, drily, 'it is left to *tumble* down, and the sheep have to eat the rubbish!'

The shepherd attributed all the changes to the death of the old Sussex farmers. He compared several well-known men with the farmers of the past. 'They have money,' he remarked, 'and they have so many acres, but they know no more about real farming than my dog does.' He reeled off names of many old farmers, including John Brown of West Blatchington, Richard Brown of Lewes, old Mr. Brown of Friston Place, Mr. Hart of Beddingham, Mr. Saxby of West Firle, and Mr. Madgwick of Alciston. Like many more old shepherds he has a wide knowledge of the farms of the district and the changes that have occurred.

We spoke of umbrellas, but he does not consider them of much use on East Dean hills owing to strong winds. He said: 'A good coat is better, but we can't get the old white overcoats now like we used to buy at Lewes. *They* were the things to keep you dry! The first day I had one the rain fell all day long, but when I took it off I was dry, although the coat was a rare weight with the wet.

'Those coats used to be made by a woman,' he continued, 'and the seams were not machined like they were in the last one I had. She used two needles and made a very strong watertight seam.' The price of a white coat was about 32s. 6d. Some farmers provided one for their shepherd so that he should not be prevented from tending or doing what was necessary on rainy days. At that time he also used to wear the round black hats, known as 'parsons' hats' or 'chum-meys,' which he also bought at Lewes. Duly was wearing 'false tongues' on his boots. His leggings were lying with his dinner-bag.

Mr. Duly has a love for canister bells. He had a 'ring' of twenty-one, but it is now incomplete. 'I sold one to an artist,' he said. 'I didn't want to, but he was so persuasive that at last I let him have one. After that I lost some among thick heather, and that spoilt the ring.'

My reference to the two dogs in the fold led the shepherd to chat about them. He prefers bob-tails or rough-haired dogs to collies as he has found them better for work. He said that where the feeding-ground is bordered by crops a bob-tail will watch and see that the sheep do not intrude over the border line, but he has not found that collies do this so well. Bob-tails may feel the heat as much as most dogs, but they disregard cold and wet and work through it, whereas a collie will endeavour to take shelter from drenching rain under the shepherd's big coat or some other cover.

The shepherd's leisure time came to an end. My long return journey did not allow a wait until evening for another chat and so we parted.

GEORGE BAILEY

Miss I.A. Battye, of Kensington, sent me a little picture of a shepherd whom she met in the Steyning district. I had not met him, so took the photo to the old shop in Steyning where horn lanterns and old-fashioned things were once obtained by some of the Sussex shepherds. Mr. Rice of the shop at once said: 'That is George Bailey of Beeding Court Farm,' and so I picked up the clue I wanted.

I found that Mr. Bailey had retired from regular work, the flock being in charge of his nephew, Mr. Bazen. He was out when I called to see him and I arranged to go again after a ramble, but after climbing a steep lane I chanced to hear the voices of sheep, and on mounting a bank saw Mr. Bailey and his nephew in a fold. They posed for their portraits, and later I followed Mr. Bailey to his home for a chat.

The old shepherd was born in 1856. He arrived in Sussex from Wiltshire when only three years old, and after moving about here and there with his father (also a shepherd), settled at Myrtle Grove for some time. From there he came to Beeding Court Farm, where he stayed for forty-six years. Though no longer 'the shepherd,' he still drifts to the fold, as one might expect, for the sheep are his chief interest.

Mr. Bailey has vivid recollections of the old harvest suppers, shearings, fairs, and other notable events. He looked forward to the fairs. 'There were three "lamb fairs,"' he said, 'St. John's, Findon (July), and Horsham, and three big fairs, Lindfield, Findon (September), and Lewes.

'The old shearing gangs were great boys,' he remarked.

George Bailey, shepherd at Beeding Court Farm for forty-six years

'There was the Bury gang, the Steyning gang, the Fulking gang and the Clapham gang. Their numbers were not always exactly the same, but usually about sixteen or eighteen. Some were very fond of their beer, and there was some rough play among them at times, but still they were great boys, and I have had plenty of fun watching them when they came at shearing.'

I asked Mr. Bailey about smocks. He always used to wear them, and bought them at Lewes. His last one came to an untimely end. He had a big fire-place in the room used as a temporary home in lambing time. His mate, before leaving, made up a big fire and left it for him, but in the meantime the flames caught more than they were supposed to. Much was destroyed, including his smock.

The shepherd gave me an old black umbrella, bought at Lewes. He has had green ones, but those came from Storrington. He still has his shears in an old pouch. The latter is an exceptional specimen; the thick leather point has a farthing on each side held by a rivet.

Mr. Bailey has followed his father's ways, and is skilled in animal cutting. I saw his flat metal stretcher and two searing irons. They have had a lot of use. Only the night previous to my visit a farmer from a distance had asked for his services, but he does not feel that he can tackle big animals any more. I appreciated the shepherd's remarks. He reminded me that the rearing of domestic animals brings the necessity for a number of rather cruel operations, and that, because it is not a cheerful subject for discussion or comment in the ordinary way, it is not generally remembered. His sympathy for the animals is strong, and he is content to know that his skill in dealing with them has saved many from prolonged and unnecessary pain.

'A shepherd's life,' said Mr. Bailey, 'is full of anxiety. You never know what may happen in the few hours you may be away from the sheep. They may be frightened or they may break loose, find wrong feed, get blown out and die before you go back! I have had all sorts of experiences,' he said, 'but the worst thing of all to a shepherd is to be with a farmer who will not feed the flock. Once, long ago, I called a farmer's attention to some sheep and told him they wanted hay. "Hay!" he said, "'tis only November! – they don't want hay; they're all right!" So there 'twas! Well, I went on for another week, and then I said: "You'll lose some if you don't send some hay up." He came to see them again. "My! they look bad," he said. "You shall have some hay." After a few more days he sent a little, but by then several had died.'

'It would save such a lot of bother,' said Mr. Bailey, 'if farmers would allow their shepherds to know what is best for the sheep.' His remark was simply and quietly made, but his gentleness only seemed to emphasize the sincerity which prompted it. It seemed a matter for regret that such an observation was considered necessary by one who had spent his life in the service of Sussex flocks.

CHARLES FUNNELL

It was on a September day that I went on a visit to the Long Man at Wilmington.[13] Seated on a roadside culvert, with only a snail for company, I ate my lunch and watched a sparrow-hawk which hovered for a long time above a hedgerow opposite, then I passed along the quaint little village street and up the hill to the church and the old priory ruins. Close by I found the Long Man, who proved so much less imposing than he had done from a distance that I soon put a hundred yards between us and turned my attention to some sheep feeding on the hillside, for a little bird had told me that a flock thereabouts wore wide-mouthed brass bells,

Charles Funnell. Barclay Wills' picture of the popular Wilmington shepherd as he was in the late 1920s: 'A most entertaining companion, full of quiet humour and quaint philosophy.'

sometimes known by the misleading names of 'brass canisters' and 'brass cluckets'.

No merry jingle rewarded me for my walk, and I was about to turn away when I spied the shepherd, crook in hand, coming towards me, and I knew from my friend's excellent description that the shepherd was Mr. Charlie Funnell. Having introduced myself and been accepted as 'a friend of a friend', we wandered after the sheep and chatted.

I found Mr. Funnell (known to most people as 'Darkie') a most entertaining companion, full of quiet humour and quaint philosophy. Although sixty-three years of age he has never been away from Lewes district. Permission to photograph him was readily accepted. He is quite used to posing for his portrait,[14] as many people visit the locality in summer to see the 'Long Man'. 'They like to take me,' he said, 'because they think I'm old-fashioned! I don't think I'm a bit old-fashioned,' he continued, with a suspicion of a humorous twinkle in his eyes. 'I think I'm quite an up-to-date man!'

The removal of his flat black hat for a second 'snap' revealed a shock of snow-white hair, and it struck me that the shepherd's unusual and handsome head – the dark, clear-cut features, the white hair and the little gold ear-rings – was a subject for the brush of an artist rather than one for my little camera.

As it was not convenient to go home to inspect the bells that pleasure was deferred for a time, but the shepherd related the history of them and the events which had prompted his jealous care of them for many years.

I inquired as to his favourite brand of tobacco and was shown his cherished oval copper 'bacca box, with his name engraved on it. It holds a modest ounce, but is filled at least once a day, and an extra ounce for luck makes his allowance eight ounces per week. I reminded him that as he could not smoke and talk at the same time my visit saved him expense, for we found many topics of mutual interest. I was entertained by his account of an experience on the hills, when he lay down in a cold, exposed place and stayed too long. He felt a numbness creeping up his body, yet seemed unable to bring himself to move. Something inside him told him that he must try to rise, but the sheep before him seemed to be moving round and round, and he felt fixed to the ground. It was only by a great effort that he rose, and the pain afterwards, as his blood circulated, was 'something to be remembered'. He is of the opinion that to be frozen to death would be the easiest and most gentle end that anyone could have.

Other tales, less serious, passed the time away, and folding time approached. He delayed a few minutes to give me some details about the oxen of the old days, and his account of the shoeing provided me with information that was fresh to me. This was the fact that while the smith was shoeing a boy stood by holding a 'Sussex trug', to the handle of which was fastened a piece of fat pork ('Real old-fashioned pork', he said, 'with fat inches thick'). In this cushion of fat the nails were stuck, and if by chance a driven nail hurt the ox its bad effect was counteracted by the pork grease.

Before I left I enlisted the shepherd's help in finding a special bell for the Worthing Museum collection, and also asked him to remember me if he acquired any old bone lockyers. To my surprise he produced from his pocket a fine old one cut from a rib bone. It had scarcely been used at all, and for that reason was a valuable addition to my series of the usual worn and grooved specimens. I accepted it with delight, a small but interesting souvenir of my trip to see the jolly shepherd of Wilmington.

A Ramble with Nelson Coppard[15]

What a treat are those occasional wonderful winter days that help us through the weeks of uncomfortable weather! – days when sudden spells of sunshine invite us to linger in the lee of a sheltered bank or hedgerow. I had tramped to one of my favourite Downland haunts and found a warm corner among a clump of stunted thorns, furze, and rough herbage, and stood there watching a blackbird and a stonechat searching in their own particular ways for something to appease their hunger.

From a deep valley beyond the ploughed ground in front of me a muffled sound of bells floated up; but the haze, which the sun had not quite dispersed, hid the flock from sight. Presently, as the mist lifted, I suddenly spied a few sheep with the sunlight shining on them, and a moment later the shepherd came into view, waving his crook, for his keen eyes had noticed me at once.

We met among the furrows, and were soon discussing many topics. As we doddled along his big collie followed us; then, as we halted a minute, he sat down to wait. Suddenly his ears were raised and his body quivered. His imploring glance at the shepherd's face was enough. His master 'clucked' once (as one would do to start a horse) and away sped the dog over the furrows. Two eyes, keener than my own, saw a rabbit in the distance, but two ears, keener than our four put together, had heard the pitiful squeal as a stoat had pounced on his victim. 'Come on!' cried the shepherd;

his pace was wonderful[16] as he made for the rabbit, while the dog gave all his attention to his unsuccessful chase of the stoat. By the time I reached my friend he had picked up the rabbit, which was as though mesmerized. It did not struggle, but sat as a stuffed creature in his hands, its body taut, its eyes fixed. The dog returned, and seeing the rabbit, reached up and washed its face with his tongue. Still the poor dear did not move. 'She be done for,' said the shepherd, as I stroked the rabbit's head. 'Ef I putt her down, Mus' Stoat'd soon fin' her again. 'Tis often the way!' So saying, he turned away and, with one quick dexterous movement, he finished the stoat's work. 'I bean't 'lowed to take a rabbit,' he said, 'but I reckon as 'tis a pity a leave 'im for Mus' Stoat or Mas'r Fox, so I'll hide 'un for ye till ye be ready to goo.' He turned homewards shortly after, leaving me in charge of the flock while he dined.

My team were no trouble. They headed the right way and the bells kept up a merry, tuneful jingle. I placed my lunch on the turf of a little mound and stood as I ate, noting the sounds of the bells. I would rather lay my food on the thyme-scented turf than on a dirty plate, although some folks are not so fussy.

A magpie sailed by; he quickened his pace as he caught sight of a stranger in his domain. Presently I followed the shepherd home, and having reported that the flock were making for the right hill, I shared his pot of tea. Soon he picked up his coat and crook, and with the collie in attendance we crossed to some ground where he had noted a few worked flints. On the way I picked up an old ox-shoe from a rut with four nails still in it, and as we reached the flinting ground I found one good specimen.

The collie grew impatient. He looked very disgusted when his master said: 'Be quiet, ye monkey!' and was not satisfied until we moved on and reached the rough grass and bushes. Here he caught sight of his old enemy once more. 'Cluck!' went the shepherd's tongue, for he had seen the stoat too; but though the dog raced off at the command the agile creature escaped again. Then a beautiful 'longtail' rose just in front of us and sped away to some trees over the valley. ''Twur ju's here that he killed a pheasan',' said my companion. With twinkling eyes he continued: 'He diddun' know as he was wrong till I told him so an' putt the pheasan' in me pocket! T'missus stuffed 'un an' cooked 'un, an' I had half of 'un for brakfus' an' t'other half for supper, for she doan' like pheasan'. I had a reg'lar good feed thet day, so I did!'

As we doddled about, slowly following the sheep, the shepherd pointed out a track made by a fox as he ran along a furrow. 'Reckon I'd like to meet he,' he said, 'for I know a man as'd give fifteen shillin's for his jacket.' He fumbled about in his pocket. 'Here's somethin' for ye,' he remarked, and produced five fox teeth. 'I foun' two in one place an' three in another,' he told me.

As the sunlight began to wane the air turned chilly. ''Tis time to turn 'em back,' explained the shepherd, as we altered our course, and we started to round up the odd members of the flock, who had strayed into quiet spots among thick herbage and furze-bushes. 'Better jes' give a look over this brow an' mek' sure,' said my friend; and we did so. We flushed a brace of partridges, and each one dropped a small wing feather, which floated to our feet; but no more sheep were found, so we stood for a minute viewing the scene before us.

The green track in the deep valley was deserted, but on the hill-side below the next brow were two men busy at a rabbit bury. 'There be ole Mike,' said the shepherd; 'bet ye I mek' him tark to me!' So saying he put his big palm to the side of his mouth and shouted: 'Putt ye head down t'hole, an' stop 'em frum comin' out!' The big voice reverberated in the still air. Instantly the reply came across the two hundred yards stretch like a clear distant voice on a telephone: 'I c'd do thet an' all. I bean't like some, got 'ead as 'ud fill a bucket!'

The rabbit-catcher must have heard our burst of laughter. The shepherd shook as he chuckled. 'I telled ye, diddun I?' he gasped. 'I telled ye! I ketch 'un every time! Reckon all they rabbits be reg'lar froughten' wi' him shoutin', but he can't kip quiet ef you speaks first, an' his tongue is allus ready wi' a good answer for ye!' His laughter bubbled out again and again, and lasted until we reached the fold. Here I left the lonely couple who work together for the good of the flock, who converse by secret signs and understand one

another so well. 'Allus pleased to see ye,' said the shepherd as we parted; and as if to endorse his statement the collie thrust his nose into my hand. I stopped by a stack to pick up a long, slim bundle. I disposed of it in the correct manner, and nobody suspected from my appearance that I had a bunny for a companion.

My friend has a hard life, and is the victim of every kind of cruel weather; yet inside him is that germ which can only flourish in the open air. Every mark and track is a message to him. The rustle of a bolting rabbit, the swish of a flushed game-bird's wings, and the song of his treasured bells mean so much to him that he needs no newspaper to amuse him. I believe that if he were confined among bricks and mortar in the town he would soon pine away.

— THE SHEPHERD'S YEAR —

Lambing

LANTERN LIGHT

My friend the shepherd[1] had arranged his sheep-fold in a little hollow among the hills of Falmer, and thither I trudged to spend a few hours with him and to share, as far as possible, the duties of the day.

As 'shepherd's mate' I am afraid I was like the shepherd boy in Wordsworth's poem[2] – something between a hindrance and a help – but I accepted my friend's invitation to extend my visit and spend the night with him.

Without the sunshine of the previous day the north-east wind seemed doubly keen. A weakly lamb had arrived during the afternoon. 'Reckon it must be fed,' observed the shepherd, 'or it'll be a deader,' so after making all secure we walked to the farm, about half a mile away. Luckily there had been no bottle babies during the first week, but now this little extra duty had begun.

It was an easy walk over the soft Downland turf.[3] The hills gradually faded. Masses of furze became dark, myster-ious lumps of shadow, which swallowed up the dog, as he hunted in vain for skulking rabbits. At length the farm loomed up in front of us and it was not long before I had shouldered a basket containing bottles of milk and some provisions, the shepherd carrying other tackle. Our return journey was made by moonlight. As we neared the gate leading towards the fold we turned off past the badger's home and cut across under the beeches with uncertain steps, for the black shadows of the branches chopped up the ground into all sorts of shapes, which appeared like roots and holes, so that it was almost a relief to step away from the queer zig-zag pattern and enter the valley where the flock lay.

Inside the shepherd's hut the little stove gave out a welcome warmth, for there was a touch of frost in the air. The milk was soon warmed and bottled, and the weak baby recovered amazingly as the magic warm fluid was sucked down.

Back to the hut we went and I was shown the two kinds of teats for the milk bottle – a modern one of rubber, and an old pattern as used by the shepherd's father – a cork with a groove at the side which held a pipe of elder twig bound in with rag.[4] From the window we looked down on the hundred and thirty-seven waiting ewes in the big pen outside, all lying contentedly among the straw litter.

Shepherd's gear and treasures

'At lambing time, visitors may find one part arranged as a sleeping bunk. By the door one corner is occupied by a small stove, with kettle, milk saucepan and firewood; the other by broom and crook. Somewhere there will be crocks for meal-times, a feeding bottle for lambs, and other receptacles, whose mysterious contents, so useful in the lambing-fold, do not always agree with the labels that still adhere to them. The lantern is an important item: in the daytime it hangs on its nail, but at night it stands ready for instant use.'

For the first few hours the moonlight was bright, and one felt that the lantern was hardly necessary for our visits to the fold. As we made our tours of inspection many of the ewes rose or moved and the bells rang loudly in the still night air. All the lambs there were born to the sound of the sheep bells.

The old man has a whimsical humour which never fails. He affected to treat me as a shepherd boy. 'Now, boy,' he said, 'hold thet light steady, can't ye, an' doan't jig 'un 'bout, or they ewes 'll be froughten.' I obeyed meekly and got my reward. 'Hap I'll mek' a shepherd of ye yet!' he said; 'put thet kettle on if ye want to, for it be turnin' colder.' As we sipped our tea we still heard an occasional rattle from a 'cluck' bell, or the loud tinkle from the one 'latten' bell, and once, a low growl from the dog on guard outside.

We snatched an hour's sleep and woke at three o'clock. The moonlight was gone and only the stars kept vigil. Fortunately the piercing north-east wind had dropped a little, for we had to spend some time with a ewe before her baby arrived. Two more were found and moved to pens.

Four o'clock came. The shepherd was asleep on a bag of pollard, while I sat on another and wrote my notes by lantern light. At four-thirty he awoke. With mock solemnity he twitted me. Had I been round the fold? No! Had I put milk on to warm? No! Then what was the use of a mate at all? On went our coats again and we made the round – an easy tour this time, just one more baby, strong and well. My friend took the lantern and went for the milk, and once more I feasted my eyes on the scene, for the old horn-windowed lantern is a thing of beauty when in use. I stood behind and watched the shepherd simply for my own pleasure, for as the big, cloaked figure ascended the wooden steps of the hut I saw some fanciful resemblance to the figure of 'Father Christmas.' Soon he returned and once more the mellow, pale orange light flickered over the dusky forms on the straw as we fed the weakly baby again.

Five o'clock at last! No need for the lantern in the fold now, for the light of dawn was changing the sky. ''Tis brakfus' time,' said the shepherd. So I went outside to fill our kettle from the churn and I did the job quickly, for the wind was back again and a thin coat of frost had appeared on the grass, the straw, and the wool of the resting sheep.

Two hours' work after breakfast was very welcome. Pens were rearranged and cleaned, the ewes were let out for a stretch and a feed and the fold tidied. The ice in the water trough was broken and fresh water pumped in. A quick inspection of the ewes and babies on the hill-side was made and the count found to be correct. I was mildly surprised when the shepherd said, 'Reckon 'tis brakfus' time,' but found this to be part of the usual routine. So on went the old kettle once more and, after a quick cold wash in a bucket, we again had breakfast. Crocks were cleaned, the hut swept, the fire raked out, and the day's round began again. We drove nearly a hundred and thirty ewes to the dewpond and back. Many were called to order by name. 'Granny' was last of all. She could not hurry. The shepherd has a soft spot for poor 'Granny.' His advice had been disregarded and she was not killed. Now the poor barren ewe, often ailing, plods along after the rest.

The time of my departure approached. I packed my camera and an old 'rumbler' cattle bell, which hung in the hut. This was very worn, and my offer to replace it by one in very good condition had been accepted.

I left my friend forking mangolds into a trug, and trudged over the hills. That night, as I sank into a comfortable feather bed, I thought of him on his sack of pollard!

LAMBING TIME

As lambing-time approaches, the site for the fold is chosen. Sometimes it is built near a rick, so that the necessary supply of hay or litter can be procured easily. The coops, placed in rows, hold the sweet-scented hay, and the sheep stand all along, thrusting their heads between the wooden bars to enjoy the meal provided. The shepherd's hut is placed in position near, for he must be on duty day and night, with short intervals, from the time the first lamb arrives. With his hand-lamp (which has now taken the place of the old-fashioned lanthorn in most places) he makes his round during the dark hours, giving any assistance he can. Should any ewes die, or be unable to feed their babies, the shepherd is saddled with further responsibility. His 'bottle babies' give him many an extra journey for their daily ration, but afterwards he always retains their friendship. Occasionally he has been able to transfer an orphan to a ewe that had lost her baby. The skin of the dead lamb is tied on the 'foster-child' and the experiment is successful.

Bells are often taken off at this time, but quiet ewes, that do not mind the constant visits of the shepherd with his lamp in the night, retain them. So it happens that some babies are born to the sound of a bell, and in these cases they display a wonderful memory for the note. The shepherd has experimented by taking the bell away. The lamb then could not locate its mother, but came to him instead when he sounded the bell.

There is still time, at this date, to welcome the arrival of many more tiny lambs. Make an effort to see them, for their real babyhood will soon pass. The spring blossoms will linger with us and will be replaced gradually by other varieties, but by that time our woolly pets will be quite large, and when the glorious Chalk-hill Blue butterfly[5]

appears on the Downs the ewes and their children will be parted. Some of the lambs will be sold, and certain of those retained will be fitted with a bell, according to the shepherd's fancy. At first there may be a scamper round, but soon the sound grows familiar. Where the old canister bells are still used, the smallest ones are often kept for these lambs (known as *tegs*). As they grow to maturity they can carry a larger bell, and it may be that, as we wander near the flock again at a subsequent date, the much-admired babies of this spring will be cheering us with their music.

An hour with the lambs on the Downs is well spent, for everything there is soothing and refreshing. The old familiar flowers give us their yearly greeting. The stonechat has donned his black head-dress and snow-white collar. (The shepherd calls him 'Furze Jack,' and says that he was married on St. Valentine's Day). From his perch on the topmost spray of golden furze he calls to his wife with quick and curious note.

A DAY AT A DOWNLAND LAMBING-FOLD

My way towards the fold led me through the farm path of a large estate. It was a dirty, rutty path, with so many traps for the feet that it was difficult to take full notice of everything inside the low hedge which bordered a newly thrown coppice.

A gap suddenly offered me an entrance. A chiff-chaff was busy on the ground and a thrush on a stump was singing. He seemed to say, 'Come inside, come inside!' and the clumps of wild arum and dog's mercury looked so cool and fresh that I responded to the invitation. I dawdled among them and wandered along, and in a few moments the

A lambing-fold, photographed by George Garland at Sullington, between Washington and Storrington. The lambing pens are covered with thatched hurdles and circular wattle feeding cages containing hay are in the fold

delightful fragrance of scented violets was wafted to my nose. A few odd roots were found first; then a veritable carpet of them appeared. Stumps, faggots, and logs abounded, but moss and dead leaves filled every other available space, and among them the sweet blossoms nestled. The fold seemed far away; the shepherd was waiting, yet half an hour slipped by. I left the violets regretfully (at a spot where a notice warned trespassers to beware) and found the pathway much improved. In place of the hedge and the farm track was an avenue of elms and beeches – old trees whose boles offered every attraction to the tree-creeper, which I watched with great delight as it ascended in jerks and inspected each little crevice. The avenue terminated at the corner of a hillside spinney, and the way now lay through the open meadows. Before I had covered the first one I could see the Downland slopes in the distance, and, on a brow, a clump of trees, under which the fold was placed.

The springy turf of the Downland bottom was pleasant after the meadow. The gorse on the slopes seemed alive with linnets, and as I climbed the steep path to the brow a wheatear acted as pacemaker, flying off in advance and settling again and again. I did not overtake him, however, for a side track showed wheel marks and certain imprints which were sufficient to direct me, and in a few moments the winding path brought me in view of the shepherd waving a welcome.

'Fine day to bring t' people out!' he bawled cheerily as he gripped my hand; 'you be late! I bin up some time!' The old man's sly humour is very refreshing – he had been on duty since dawn.

Never have I seen a prettier fold. If the spot had been chosen for picturesque effect only it could not have been more beautiful. The front was partly screened by hurdles banked up with furze branches, but at the sides and back grew dense furze thickets, topped with a wealth of golden blossom. Behind the furze at the back towered a row of trees, and from their tangle of branches, through which the sky showed as a blue haze, two green woodpeckers were calling. Inside the fold hurdled pens were ranged on each side, and a portion of the littered ground was divided off for a few late ewes.

A hut on wheels stood outside the fold, for no sheltering barn was available up there on the hills, no cart shed with a cosy lean-to for the shepherd to use, as in some places; consequently this hut had been my friend's only home for some weeks. Now most of his family had arrived it was possible for him to go home to his Sunday dinner. I was therefore left in charge – a very amateur shepherd – and I enjoyed the most delightful experience.

Dry twigs and coal filled the little stove in the hut. The kettle soon sang merrily, and I made tea, while I shared my lunch with the dog on guard. Through the hut window I could overlook the fold, also the enclosure where the ewes with the younger lambs were turned out – a little stretch of short turf with thyme-covered hillocks. Here the lambs got their first knowledge of the big world and learned to play 'King of the Castle' on the mounds. After their play they rested and dozed in the sun.

The woodpeckers called again. I went into the fold with the binoculars to get a closer view, and while there a baby arrived. I was able to report to the shepherd, when he returned, that 'mother and child were doing well,' to which he agreed.

We then set out to move the main flock, which had been in a big enclosure at a little distance since early morning, so that they should be nearer at folding-time. With the little ones wanting to pry about and stray there was plenty of

A shepherd and lambs. The shepherd bottle-feeds a lamb, while others lie in front of a cleft feeding crib containing hay. Behind this can just be seen bundles of brushwood faggots for protection against wind and weather

work for two, but gently and methodically it was done, and they were soon spread on the hill-brow below the fold basking in the afternoon sunshine.

The shepherd's duties continued. More twigs and coal warmed the milk he had brought for several 'bottle babies.' The 'tet' was fitted to the bottle, and one by one the anxious, hungry children were satisfied. Some of them were so eager that they readily took my finger in their mouths and sucked heartily! Grains were taken to the enclosure and the feeding troughs were soon filled by heads, large and small. Mangolds were also taken and strewed about for the ewes. Then came the feeding of the ewes with the youngest babies, still in the pens – a bundle of hay and a mangold for each one. The next job was a rigid inspection of the babies. Two had to be helped to get their milk, two who were not getting enough were given extra from the bottle, and so on, until every one had received the necessary attention.

The shepherd was in great form, in spite of his seventy years, and sang a verse from one of his father's old ballads. The long, weary hours in the stormy weather and his struggles to care for the babies that arrived in the night, in spite of wind and rain, were now disregarded. His rounds, made with the aid of a lamp, seemed, for the moment, far away, for here was a glorious sunny day which altered everything and helped the lambs.

In the hut were a lot of old canister bells which had been removed during lambing-time, and we both enjoyed a chat while inspecting them. We talked of crooks and bells and many things while we sat and smoked, but the watch told us that the flock must soon be folded, so I packed my bag and a bell which my generous friend had given me and made my way down the hill. I looked back. The glorious sun was shining on the trees. In front of the fold the shepherd stood watching me. He waved a vigorous farewell with his shining crook and turned towards his sheep.

Then I trudged along the homeward path again, and when I came to the coppice I entered. The bell in my pocket was gagged with a handkerchief; my rubbered boots made little noise; and from the abundance of scented blossoms I took what I wished.

I had set out with high spirits in the morning; I returned at night with every wish gratified, and with treasures beyond expectation – some excellent records of my friend and his Downland family, a sheep bell, a crowd of fragrant violets and another of those many memories which shall, in the future, roll back the intervening years, whenever I wish, in defiance of Old Father Time.

Sheep-washing

The heavy work of washing sheep thoroughly in cold water a week or more previous to shearing was an additional duty in former days. Farmers were naturally eager to see their wool put on the market in the cleanest and best condition possible, as clean 'clips' fetched far better prices.

The practice has now died out in most districts. Wool staplers now prefer unwashed fleeces, as lanoline is extracted from them by a special process. This was formerly washed out of the wool and wasted in the sheep-wash.

So many farms lacked a brook with a deep pool, or any other place suitable for a wash, that many flocks were driven to appointed spots every year for the purpose. One of these

Sheep-washing. '"I remember the last I saw of it," said Nelson Coppard; "there were two men in tubs to wash the sheep, two to throw the sheep in, and another standing by with a dipping hook."' The sheep were penned before being released when they were scrubbed and dipped. The washer stood in a barrel to keep dry; even so, it was a cold wet job and many suffered from rheumatism as a result

The old days are dead; the old ways are dying out. On many farms the shepherd and his assistants now shear the flock with patent clippers worked by a machine. With these modern appliances the animals are sheared quickly and closely, and as they stand in a crowd waiting for the next victims to join them they look so naked and miserable that they are a dismal sight.

Show flocks are still sheared by hand with the old shears. An even fleece, free from ridges, is thereby ensured, and the sheep do not lose too much wool.

On farms where small flocks are kept the shepherd is expected to manage the shearing by hand in the old way with the help of an assistant.

Small flocks and show flocks therefore provide us with opportunities to watch shearing in the old style – work that is far more fascinating to the onlooker than to those employed.

We may still see the sheep waiting in the hopper, see them sheared and released, and watch the fleeces being wound into a compact bundle. Sometimes we may see a keg of beer, supplied for the shearers, in a corner of the shed. So much is still left to us, but one feature of the old shearings is gone for ever – the very attractive little figure of the tar-boy. 'Lowest in the gang, with many masters to please, expecting no praise for instant service (and a kick for anything less)' – such was the word-picture passed on to me! Instantly my sympathy and interest were centred on him. It seemed quite worth while to arrange an enlargement of the tar-boy from the photo of a shearing gang.

It was my good fortune to meet Mr. Jack Hazelgrove, who, forty years ago, acted as tar-boy to the old Clapham and Patching gang. He told me that they used to start work on the farms at 6.30 a.m., and this often made it necessary to leave home at 4 a.m. From 6.30 onwards, with the

The Tar-boy, identified by Barclay Wills as Frank Shepherd. '"Lowest in the gang, with many masters to please, expecting no praise for instant service (and a kick for anything less)"', whose job it was to dress cuts on sheep, that sometimes occurred during shearing, with a dab of tar and to bring refreshments to the shearers

exception of dinner time, he was kept busy. He explained to me that tar was not just dabbed on as one might think. It was put on with his finger and as the tar dropped on the wound the finger was turned over and the tar spread with it. At some farms tar was not used, and he was provided with wood ashes. At the call 'Tar-boy' he attended, dropped a little wood ash on the wound and spread it gently with his finger.

When intervals for refreshment were announced it was his duty to fill the shearers' mugs with beer, while in his spare minutes he helped the winder, and learned to wind fleeces properly. Later he learned to shear by beginning or finishing ewes, and became a 'colt.'

When work was over for the day the gang returned home, but occasionally where there was a large flock to shear they would take food for two days. They slept in a barn or cart-shed. After such work a farm waggon lined with fleeces was a welcome bed.

On the long journeys to and fro the most direct paths to outlying farms were used. At other odd times members of the gang would meet for a ramble, and on such occasions it was their pleasure to tread out footpaths and keep them open. By this means Mr. Hazelgrove obtained a wonderful knowledge of the paths and rights of way around the district, and he uses them still. I trod out some of them with him, and feel sure that some of the delightful walks enjoyed by us to-day would probably have been lost if the footpaths had not been trodden out and kept open by members of the old shearing gangs.

At fairs and sheep sales it is usual for farmers who are selling to treat their shepherds to some drink during the hours they stand by the pens, or to give them money for that purpose. In addition to this it was customary at some places for the head of the auctioneering firm to provide free beer to those shepherds in charge of the sheep. The reason for the abrupt stoppage of this custom at one fair ground was related to me, in confidence, as I stood chatting by a fold. It appears that D——, an East Sussex shepherd, was known as a man who enjoyed his beer, and his capacity for carrying it made others envious. With other shepherds he attended the booth for the usual free beer. It was given to them in a large jug, which was emptied and returned in due course. D—— was still thirsty. He said to the others: 'It do seem a pity they should take any o' that beer away again, boys! Us ought to be smart enough to get another jug, surelye!' and he suggested a novel plan for obtaining it. Under his direction a few of them (as much for fun as for an extra drink) disguised themselves in odd hats and garments borrowed from other members of the company; sauntered to the booth and asked for beer. They were ready to give other men's names if asked. To their surprise the beer was handed over. Their success induced others to try the same method, but unknown to them, certain eyes and tongues had been busy, with dire result. Since that day there has been no more free beer supplied by that auctioneer!

Here is an old song which was included in the programme following the 'Sheep-Shearing Song.' According to the singer it was a general favourite. Like many another old country song it may differ by a line or two from other versions (a certain amount of licence being allowed to the singers), but it is easy to believe that the shearers made the most of their opportunity to join in the simple chorus.

A DROP OF GOOD BEER[15]

I'm Roger Rough the ploughman,
A ploughman, sir, am I;
Just like my thirsty father

My throttle's always dry!
The world goes round – to me 'tis right,
With no one I interfere,
But I'll sing and work from morn till night,
And then I will drink my beer!

Chorus:
I likes a drop o' good beer, I doos,
I'm fond of a drop o' good beer, I is,
Let gentlemen fine set down to their wine
And I will stick to my beer!

There's Sally – that's my wife, sir –
Likes beer as well as me,
And seems as happy in life, sir,
As a woman could seem to be;
She minds her home, takes care of the tin –
No gossiping idlers near;
Sure as every Saturday night comes round,
Like me she drinks her beer!

There's my old dad – God bless him! –
He's now turned ninety-five;
No work could ever depress him,
He's the happiest man alive!

He's old in age, but young in health,
His heart and hand both clear!
Possessed of those he keeps good health,
But still he sticks to his beer!

Now lads, need no persuasion,
But send your glasses round,
And never fear invasion

While barley grows on our grounds.
May discord cease, and trade increase
With every coming year –
When everything's crowned, and counts all paid,
Then I'll sing and drink my beer!

In Shepherd Michael Blann's old song book I found the words of a good song in praise of beer. It was written for him by his brother and that was all he knew about it. Later Mr. Arthur Beckett informed me that most of it was taken from a collection of verses by John Hollamby of Hailsham, published in *Our Sussex Parish*[16] by Thomas Geering. A copy of the song, as I found it, is given below.

BLANN'S BEER

Oh, 'Blanns' is the beer for me;
  A pint of it's so handy,
It is as fine as any wine
  And strong as any brandy!

If you are ill 'twill make you well
  And put you in condition;
A man that will drink Blann's old ale
  Has need of no physician!

Chorus:
Oh, Blann's is the beer for me, etc.

'Twill fill your veins and warm your brains
  And drive out melancholy;
Your nerves 'twill brace, and paint your face,
  And make you fat and jolly.

The foreigners may praise their wines! –
   'Tis only to deceive us!
Would they come here and taste this beer
   I'm sure they'd never leave us!

The meagre French their thirst would quench,
   And find much good 'twould do them; –
Keep them a year on Blann's good beer,
   Their country would not know them.

All you that have not tasted it,
   I'd have you set about it;
No man with pence and common sense
   Would ever be without it!

Trimming

George Humphrey of Sompting. '"Will you leave your work for once and pose for a photograph for a shilling?" "Reckon I will!" he replied, briskly. "I'd do anything for a shillin'!" . . . the result was the best photograph ever obtained of this old Sussex shepherd.' Barclay Wills' photograph shows a sheep bow in use, a pair of shears, spiked on wattle hurdles, at the ready

Among the interesting items included in a shepherd's outfit sheep bows claim a place. These are used for holding lambs by the head while they are trimmed and made tidy for market.

They are natural branches of forked shape shaved down to the required size. Near the top of each of the two forked arms are two or three round holes. A thin iron rod on a chain is fastened to the pole near the fork by a staple or a bolt.

The pointed end of the pole is driven into the ground by blows on the crutch of the fork with an iron driver, and when a lamb's head is placed in the fork the rod is slid through the most convenient pair of holes.

Most of the bows are plain wood – some are polished in parts with use – but I have seen some which had been painted blue years ago, like the old farm wagons. I have heard them described as 'blue bows' and as 'strods.' 'Down west,' by Salisbury, they are known as 'stocks.' I own a well- preserved specimen made by Nelson Coppard, at Iford, about 1894, and others were given to me by Henry Coppard of Patcham.[17] These have hung on my wall for some months, but to-day (September 1st), for the first time, I saw some actually in use and I preferred to figure one thus, although it is very old and worn, for each year the chances of seeing this interesting sight become more rare.

Sheltered by a hedge and a row of sycamore trees I found a tiny fold at Lychpole,[18] where Mr. Bennett, the shepherd, and two assistants were trimming lambs ready for Findon Fair. At the back the finished lambs kept company with those waiting their turn. The thick carpet of straw was well trodden and three bows were in use in front, near the hurdles. Each lamb was first cleaned down with a stiff brush; then the trimming commenced. Part being done the victim was turned round, its head was put into the fork of the bows, the rod was shot, and the lamb was a prisoner until the operation was over. A piece of sacking was placed over the lamb's head to keep it quiet. As the thin layers of wool were snipped off there appeared the clean, crisp coats which gave the lambs such an attractive appearance as they stand in the pens on fair day.

Sometimes a shepherd when trimming lambs will place a cloth on the ground as at shearing time, and so save the snippets of wool. These, when washed, make excellent stuffing for cushions.

After the trimming the shears will be put away until about May, when shearing time comes round again. Then many of these same lambs will experience the loss of wool for the second time.

While watching the work I remembered the words of an old shepherd, who said to me, 'One way an' 'nother they do put up wi' summat; they be born orkard, same's we, an' they 'as to learn. Thet's why I do be kind 's I can, but sum on 'em be turr'ble orkard chil'ren, an' no mistake!'

Findon Fair [19]

From the sale catalogue I learned that ten thousand sheep and lambs were to be sold on the Fair Ground on September 14th.

Bidding was in progress when I arrived, so I took my place among those who had real interest in the sale and who crowded round the auctioneer's stool by the side of the pens.

'Will anyone start at sixty shillings?' the auctioneer was saying. 'A nice lot these, gentlemen – shall we say sixty shillings? – fifty-five! Thank you, sir, fifty-five bid – fifty-five shillings – fifty-six – seven – eight – nine – now gentlemen! Sixty? Thank you. Sixty shillings, going at sixty! – sixty-one – sixty-one bid – sixty-one and a half – sixty-two! Any more, gentlemen? Going at sixty-two – for the last time, sixty-two shillings!' Bang went the book on the palm of his hand, the deal was clinched, and another fifty dumb creatures were destined to be driven over the hills to a new home and to obey a new master and a new dog!

The bidding recommenced. In about two minutes bang went the book again and a pen of fat, panting lambs were claimed by a Worthing butcher. I took a snapshot of the auctioneer on the stool, and his assistant, who stood in each pen and prodded and tapped the occupants with a short crooked stick. The precise reason for this action was not apparent. Was it necessary, or was it only a part of the programme? No amount of tapping and prodding could change dull, pensive-looking animals as sheep into alert, brisk, intelligent-looking ones, especially when huddled together inside a little market pen. I turned away, for I had other interests. I did not go to buy sheep, anyway, for I only had twelve and sixpence.

As I walked between the long rows of pens I spied some of the various shepherds who have helped me with my study of bells and crooks. One had an old crook for me, which was a pleasant surprise. Then I found some lambs that I had last seen at Sompting when they were being trimmed. Each in turn had stood with its head held in the old 'bows' under a piece of sacking, while it was tidied up for the fair. 'I shan't want one of the bows any more,' said the shepherd (who is moving at Michaelmas), 'so if you would like to have them you'll know where to find them,' and he arranged a hiding-place under a corn stack so that I might go at my leisure to pick up the treasure; a pleasant surprise again. Next I met a man who used a very peculiar crook last fair day. He had it with him again, but I did not recognize it, for in the interval it had snapped and had been repaired by a smith. It was still a useful crook, but the beauty of its curves had departed. I was not surprised to learn that the owner was dissatisfied with it and had arranged to have a new one made to his own design.

The day was very warm and the animals looked tired, penned up so closely. So did the dogs, which were tied, at intervals, to tent pegs, hurdles and wheels, but nobody seemed to think anything of that. So far as I could see there was no provision made for giving any animal a drink. Some may say that it was not necessary (probably because it was not easy to arrange), but one thoughtful man hung a few cabbages round his pen and the bare stalks soon bore witness to the animal's appreciation. It seemed a pity that the idea was not generally adopted. If drink could not be provided a little greenstuff could and should have been given. Farmers were getting money for their stocks; they would never see the animals again; what did they care, once the auctioneer had banged his book on his hand? The answer is 'Nothing!' The fact is that only the shepherds care, but they have no power. Their careful nursing, watching, feeding and doctoring brings the sheep into condition and puts a fair profit into their employers' pockets, for their own pittance does not take much gilt off the gross total.

The roundabouts and coco-nut shies were busy and the refreshment tent was filled with clients, for the same warm sun that tried the sheep brought a number of visitors to the fair ground. It also made it possible for a certain old shepherd to attend. He seemed less deaf and more animated than usual. Perhaps the surroundings had some subtle effect on his system. He told me of the fairs years ago, of old Sussex farmers, jovial and generous, who sold their own sheep, of the old 'round frocks,' and much more. He told me of Berry's ancient Pyecombe crooks. 'Look out fur one wid a B at de top o' de guide,' he said, 'fur dey be older'n any.' He made some very caustic comments on the presence of so many motors, and deplored the changes in ways and habits. He mentioned another shepherd. 'What sort o' flock has he got now then?' he asked. 'Between two and three hundred,' I told him. 'Ah!' he said, 'a nice cosy li'l' ole

Findon Fair, photographed by George Garland in the 1930s. Many of the sheep would have been 'driven' to the fair, despite the increasing use of motor transport evident. Horace Oliver, a Burpham shepherd, used to get up at two in the morning in order to 'drive' his sheep to Findon by six. The wattle house where the hurdles are stored is on the extreme left, with the fair ground and gipsy caravans beyond. The fair, held on the second Saturday of September, was an annual pilgrimage for the Downland shepherds, a time to meet old friends and swap stories

place; reckon I c'd jus' manage a job like thet.' What pluck for an octogenarian.

Truly times have changed, as he said, as the ugly car park and a hovering aeroplane showed, for a Red Admiral butterfly, which danced gaily over the pens, was the only thing which looked exactly the same to the aged shepherd!

The Shepherd's Christmas

Few people expect to go to their ordinary work on Christmas Day, and fewer still on every one of the other three hundred and sixty-four days also; yet this is the record of a certain Downland shepherd, and he does it without grumbling.

He will not be at home for a merry dinner-party at midday, but his grandchildren will welcome him at tea-time when his work is done. A week ago he said to me: 'Come Christmas Day I'll be right over t' brow yonder. Goo home to dinner? No, I wun't; 'tis too far there an' back agin. Reckon I sh'd be middlin' hungry time I got back to team, an' want 'nother dinner! I bean't so young as I was. One time I ran up an' down t' old hills like doin' nothing', but now I goes slower. I be like t' old bell yokes – they doos their work an' they be tough, but they wears an' wears till they breaks asunder – an' I be wearin'!'

Should the day be fine my friend will eat his frugal lunch in the open, with his dog for company, and the sheep bells will provide the music he loves best. Last Christmas, to my surprise, he made me a present of his favourite big bell.[20] Pleased as I was to accept it, the fact remains that a bell was the only thing he had to give, and he gave the best he owned. How few of us would willingly offer our most cherished possession to a friend as a Christmas gift!

The shepherd gets no stone of beef at Christmas as he did in the old days when Sussex farms were worked by Sussex farmers. He can recall his enjoyment of these days – of harvest suppers, of little remembrances at lambing time and shearing time.

Sometimes, in course of conversation, I obtain a glimpse of the sturdy, independent character of the old Downsman. He is a man of simple faith, who would stand against a hundred others for his opinion, and would work like a slave from a sense of duty; a man who will give away his favourite possession, who will point out the beauty of a wild flower, and go to any trouble to prevent one of his flock from wearing an uncomfortable bell-collar. Perhaps his life of unending work among the beauties of Nature has caused him to view life from a different standpoint from ours, but his quaintly expressed opinions are often equal to the cream of a church sermon.

Should you have a spare hour or two at this season of peace and goodwill, and you know of a shepherd of the old school, it will refresh you to tread the Downland paths, where peace reigns, and to hear the gentle song of the old bells. Before you start, remember that the figure with the crook is very human, and put some tobacco in your pocket for the old man on the hill.

the place of tailing irons. On one farm at Sompting it was found that for some unknown reason tails did not heal well until a change was made and a knife used. This fact leads to further consideration of the shepherd's knife, the use of which I have already described as follows:

Few people would think of the shepherd's big clasp-knife as an important tool, but it is in fact one of the most essential items in his outfit. Among its many uses I have seen it take the place of a surgeon's lancet at lambing time. With it the shepherd cuts and trims sticks of all kinds, including crook sticks; cuts out strap tackle, yokes and pegs, pares the hooves of sheep, and cuts up his lunch; while during any day a dozen other odd things may provide other uses for it. The shepherd has to rely on his own efforts to such an extent that without a strong, sharp knife, very little could be done.

The ruddle-pot, though only used at pairing time, is an important item to the shepherd. Rams are freely coloured with the ruddle, or red chalk, and the transfer of the colour to the ewes tells the shepherd what he must know ere he can count the days on the calendar and arrange for the preparation of his lambing fold at the correct date.

The fly-powder tin is one of a crowd of oddments and may be regarded as representing all the range of mysterious pots, tins and bottles which comprise the 'medicine cupboard' of the shepherd.

Two other items complete the ordinary list of shepherd's gear – the umbrella and the bag which the shepherd carries to and fro each day. It is of no set pattern – simply a bag or satchel to contain his dinner, a tea bottle, and any small thing which may be required. It is usually slung over the shoulder by a strap held in the hand, or hung over the

Shepherd Newell and his dog. He is carrying a shepherd's stool. The stool was made of a small oak board with a hole in the centre into which a short stick was fitted. His dog, Prince, is a collie, a breed generally preferred by ther Downland shepherds at this date

shoulder on the crook. I have seen various quaint things come out of some of these bags at close of day – 'shepherd's crowns,' ox shoes, flints, rabbits, bells and tackle, wires, split pins and string, and once some flowering bulbs or 'Star of Bethlehem' found on the Downs. 'Reckon 'tis a pretty flower!' the shepherd said; 'I be takin' 'em for my garden.'

Among some gear at Falmer, belonging to Nelson Coppard, was a dipping hook, once used at sheep washings. As this custom is now almost out of date the hook will soon be regarded as a curio. It was mounted on a stout pole about eight feet long. With it a sheep's head could be ducked under the water or help up as required.

On only two occasions I have found a shepherd's stool in use. This is a home-made contrivance – a small piece of strong board with a hole in the centre cut and shaped to receive the end of a short stick. This very useful 'extra' to a shepherd's outfit enables the owner to rest for a few minutes while tending when the ground is damp. My specimen is made of oak, and was given to me by Mr. G. Newell whom I met with his flock at Devil's Dyke. He had used it for twenty years. A leather brace end nailed to the underside enabled him to attach it to any convenient button, and he carried the stick separately.

From these notes it will be seen that the average shepherd's gear makes an important list, surprising to those who learn the number of his necessary possessions for the first time. Although many of the articles are quite ordinary, the wooden bows, stool, dirt-knocker, and thatching-needle are of particular interest, for with the carved bell-tackle, described later, they form a series of objects made by the shepherds for their own use.

Shepherds' Umbrellas

To meet a shepherd carrying one of the large old-fashioned umbrellas is already unusual, and such a sight will become worthy of remark by ramblers, for although these useful articles were once part of every shepherd's equipment, they are now out of date.

It is only natural that in course of time each specimen must come to a sudden end through some mishap. Violent gusts of wind, such as we all experience occasionally, have been responsible for the loss of the majority of big umbrellas.

Since war-time it has been almost useless to order one of the old type or to try to get an old frame covered with the original green material. Some modern ones obtained to order for shepherds have proved to be nothing more than an 'out size' of the ordinary black-covered umbrella, and have provoked hearty curses from those who parted with a fair sum and sent cash with order, in the hope that they would obtain what they required.

Casual mention of big umbrellas to any old shepherd will usually draw some reply in priase of them. Even those who have a well-worn specimen and do not carry it for fear of being considered 'too old-fashioned' by employer and others are willing to admit that, like the old round frocks, 'there is nothing to compare with them.' A shepherd provided with one, and clad in his thick overcoat, had only to push his way, backwards, into a bush, crouch down and hold the big umbrella over him, and he was protected from the heaviest rain. It was usual to carry one to and fro in stormy weather, and although big and cumbersome, they were easily man-

George Humphrey under his umbrella – 'A shepherd provided with one had only to push his way, backwards, into a bush, crouch down and hold the big umbrella over him, and he was protected from the heaviest rain.'

aged by attaching a cord to the handle and the ferrule end. The cord, adjusted to the proper length, was put over the head and under one arm so that the umbrella was slung across the shepherd's back. I have an old specimen which measures four feet across when opened and which weighs three pounds, yet, carried in this way, its size is scarcely noticed.

Some umbrellas had ribs of cane and some of whalebone. Many frames of the latter kind, when done with, have found their way to some dealer's store, where a good price is asked for the whalebone ribs.

When interviewing Michael Blann I spoke of big umbrellas, and he said: 'Once I had one with me when I walked miles home from Lewes in a storm of rain and wind. A gust got under it and blew it to bits, so I threw it in the ditch. Afterwards I bought another one and was caught in another storm. I was on top of a bank. The wind came very suddenly and lifted the umbrella and me with it and dropped us down under none the worse. That was a storm, and no mistake! Another gust came and blew the umbrella nearly inside out, and as I turned to face the wind it blew back to its proper shape again, but before I had time to think much of these wonderful antics a still more violent wind shattered him altogether.'

Many somewhat similar incidents are remembered by other old shepherds, but the most entertaining one I have heard was recorded in my *Bypaths in Downland* and is reprinted here:

One day I was making inquiries from a shepherd regarding these big umbrellas, and he related how he once got some fun out of the use of 'the big 'un.' The flocks of three adjoining farms were each in the charge of a young shepherd, and he was one of the three. They would sometimes meet at one particular spot for a chat, and on one occasion were caught in a sudden heavy shower. Not far from where they stood a man and a maid were sitting on a bank, and one of the shepherds suggested that my friend should take his big umbrella and share it with the couple. On the spur of the moment he accepted the challenge and walked across to them with the big umbrella open. To his amazement the girl's companion rose and ran away as fast as he could, leaving her on the bank. He evidently suspected from the young shepherd's advance and the merriment of the others that some plot was afoot, and, like the coward he was, left the girl to her fate. It would seem that he had no particular claim to her affection and that she was inclined to appreciate the young shepherd's action, for she sat quite still and allowed him to shelter her from the rain in gallant fashion. 'I diddun' mine they cheps grinnin','' he told me, 'fur she wur a purty gel, an' tarked to I. "Doan't 'ee putt t' umbereller down," she says, when t' rain stopped; "kip 'un up a bit, fur I likes bein' 'ere 'long o' you." But I *did* putt 'un down, fur t' dog telled I to.' The old shepherd chuckled and rubbed his head. 'Damme, I'd nigh forgot 'bout it all,' he said, 'till you comed tarkin' 'bout they umberellers!'

Horn Lanterns

The peculiar fascination which clings to objects connected with past days has already extended to the old horn lanterns which were commonly used by Sussex shepherds at lambing time.

As I look back on all the hours spent in the company of shepherds, under all sorts of conditions, I cannot recall a more fascinating sight than a certain shepherd, while on his round in a lambing fold at night, carrying a horn-windowed candle-lantern. Until I stood there, a silent watcher in the shadow, and experienced those moments of pure delight, I had not realized that any artificial light could hold such charm. At first I had carried the lantern myself, stepping softly among the resting ewes, and holding it while the shepherd performed the necessary services, but later I stood back while he went round alone, simply for the pleasure of gazing at the scene. The cloaked figure with the lantern seemed to glide among the sheep; the pale, mellow light flickered over his garments, over recumbent forms in the straw, and over the hurdle-pens; then, as the lantern was set down by his feet, like a great glow-worm, the yellow rays lit up the shepherd's features, and shone upon a newly born baby.

As my friend returned to the hut I drew further back, to see him ascend the steps. The door opened; the lighted interior swallowed up the lantern's rays, and the big figure was silhouetted for a moment in the doorway. He looked round for me. 'Oh, there you be, then,' he said, as I followed him in, amused that I should have been watching him at his work. Little he knew of the beauty I had found in the homely scene transformed by the magic of lantern-light!

With the possession of a fine horn lantern, used by shepherd Jesse Moulding of Goring, my interest increased. Variation in size and design of other specimens suggested that more might be learned about them. From the oldest country shops where they were once sold I was referred to wholesale dealers, who, in turn, introduced me to the actual makers. Messrs. Walker & Loach of Birmingham, who specialized in the manufacture of these lanterns, kindly furnished me with some details about them, and sent me an old price list from which I note that horn lanterns were once made in eight sizes, as follows:

	Height at Shoulder	Diameter
No. 1	6 in.	4½ in.
2	6½ "	5 "
3	7 "	5½ "
4	8 "	6 "
5	8½ "	6½ "
6	9¼ "	7¼ "
7	10 "	8 "
8	12 "	8¾ "

Two types of lantern were made. The one termed 'Plain' has the crown pierced by three sets of holes – one large, surrounded by six small, in each set. (This type, if lighted indoors, throws a pretty pattern of twenty-one spots of light on the ceiling.) The other type, termed 'Dormered,' has the holes hidden by three little covers like dormer windows, each pierced with an opening in the shape of a cross or with many small holes.

The horn windows were fitted with wire guards except in lanterns of small size.

Messrs. Walker & Loach wrote to me as follows:

Our firm, in the past, specialized in the manufacture of the horn lantern, and we have made many hundreds of grosses. Unfortunately, for some years now, the horn leaf has not been obtainable, and this has caused the trade to become obsolete.

We still make up a lantern which we term a Horn Lantern, fitted up with panes which are a substitute for the old horn leaf. They are nothing like as good, and we look upon it as a dying trade.

The firm we procured the horn leaf from was Mr. John Harmer, 68 Pope Street, Birmingham. We understand these people procured the leaf in the rough from a firm at Bewdley, near Kidderminster, and afterwards dried and polished it to the small sample we send you herewith. They gave up business about December, 1915.

The safety which results from the use of a horn lantern in a straw-littered lambing fold accounted for the old shepherds' strong preference for them. A chance knock or a kick from a ewe may have split a horn leaf sometimes, but no dreadful consequences followed.

As the last stocks in country shops sold out it became difficult for some shepherds to obtain a horn lantern, and the use of glass-windowed candle-lanterns followed. A fine heavy specimen was used by a shepherd – the late John Norris, of Coate Farm, Durrington – about sixty years ago. It may be seen in Worthing Museum.[1]

The use of modern sheets of flexible transparent composition has made it possible for anyone to acquire a lantern of the old type, for the metal part is still made to the same old pattern; but it is good to own a real horn one that was once used by a shepherd, for some subtle charm still clings to it.

We may try its effect in place of our usual indoor light when we have time to rest and day-dream. As shadows deepen, the pale rays from the candle seem to show brighter through the horn, and as the same soft glow that shone on the hurdle pens and sheep with newly born babies illuminates the room, half-forgotten scenes of folds, and shepherds, and small tottering lambs return to us. In fancy we hear once more the clamour as mothers and children call to one another. Other pictures follow, and we live past hours over again, for there is memory-magic in the light from the old horn lantern!

Shepherds' Clothes

When asked about clothes worn in past days a shepherd remarked to me: 'To write about round-frocks and corduroys and things is quite right, but I don't think you ought to put it down that every shepherd wore this or that – sometimes he did, *if he could get it!* If he couldn't, then he had to do with what he had. Shepherds' wages were very low; why, when I was saving to be married, and buying bits of furniture, I wore *anything*. For a long time I wore an old frock-coat. I didn't care what I looked like, for it saved me from spending more money. If you had an old photo of me taken at that time you might think shepherds wore frock-coats – and you would be wrong!'

This point of view was confirmed by another man whom I interviewed. He said: 'When I left home, over sixty years ago, I walked miles to get to Brighton, to save my bit of money. There was some delay before I got work and I had to sell my dog to get enough to keep me till I started. My wages were eleven shillings a week and I wore out my clothes faster than I could buy new ones. I wore anything I could get, and it was pretty rough sometimes on the hills without very much clothing!'

Many shepherds of to-day, although receiving better

Shepherd Turner of Westmeston, near Ditchling. Mr Turner's clothing is typical of the Downland shepherd. The wintry scene shows that the photograph is taken in a lambing fold. He is wearing a hard felt hat and to keep out the cold a great coat and cape made of calico or canvas waterproofed in boiled oil

wages, are still obliged to wear what they can get instead of what they would choose. It does not seem possible to obtain the old types of hard-wearing materials in suits or overcoats at ordinary prices: the corduroy has no lasting quality; the overcoats will not stand a day's rain. The consequence is that the modern shepherd is generally badly dressed in a poor corduroy or a shoddy suit, and old overcoats of shoddy material or ugly khaki army coats have taken the place of the big fleecy coats and the second-hand cavalry cloaks which were worn by some old men and which gave them such an imposing and picturesque appearance.

Old leather leggings are still seen, for they were frequently handed down or passed on from one shepherd to another. Strong boots and leggings are important items in every shepherd's outfit. The short leggings – so useful on a dewy morning – are sometimes removed at the fold if the moisture dries off, and are carried home at night. In the picture of Tom Rusbridge[2] an interesting item still worn by some shepherds is seen: 'false tongues' or large pieces of thick leather which are tied over the laced fronts of boots to keep out the dew and rain. Mr. Rusbridge offered to make me a pair as specimens, and was amused to know that I preferred to have the actual ones he wore, which are shown in the photo.

One small point is worth notice in connection with shepherds' dress: the fact that when at work the men usually have the neck and throat bare and free to the air.

In the chapter on Michael Blann there is a reference to hats of such tough material that they would support a man's weight. Subsequently a chance question on the subject brought a caustic comment from a younger man. 'You mustn't believe all those ol' fellers tell you,' he said, 'some of 'em can mek up a good yarn if they get anybody to listen!' But proof was waiting for me in *Sussex Folk and Sussex*

Tom Rusbridge of Findon. Barclay Wills' photograph was taken after Tom had been forced to retire because of rheumatism. He is wearing 'false tongues', made of thick leather, over his boots to keep out the dew and rain. A pair which belonged to him are now in Worthing Museum

Ways, by the Revd John Coker Egerton, Rector of Burwash, who died in 1888, aged fifty-eight. The following account of home-made linens and of a village hatter who made these excellent hats is taken from that book:

Within the recollection of many persons still alive we grew flax, bleached it, carded it, spun and wove it at home. In many of our cottages there are yet to be found sheets, tablecloths, and other articles of linen which seems to defy the power of time. Doubtless they are now kept more as curiosities than for use; still they have borne an amount of wear and tear which is certainly not expected of more modern goods.

We had our own hatter within my own memory, though when I knew him he had ceased to work at his trade. His productions had the character of being everlasting. It was said to be simply impossible to wear them out. One particular kind of hat, called 'dog's-hair' hats, had this further peculiarity, that if a man wished to reach something, say from a shelf, and found himself hardly tall enough, he had nothing to do but to put down his hat upon the ground and stand upon it; it would bear him without a sign of yielding. A man who used to wear one of these imperishable helmets told me that till it got well sweated to the shape of the head, wearing it was 'all one as if you had your head in the stocks.' The two finer kinds of material used in our hats were 'hare's flick' or 'rabbit flick.' Hats of the former kind were, I believe, expensive and quite aristocratic, and were reserved principally for Sundays and special occasions.

Shepherd Tom Rusbridge told me that, when he was young, shepherds and farmers wore hard felt hats with flat crowns. The shepherds painted theirs, mostly grey, and when they were done they were shiny and would keep out any wet. Dick Flint, a Findon shepherd, referred to these as 'half-high' hats.

The last time I saw shepherd Frank Upton he was wearing a stout old white smock, which had been shortened and opened down the front for convenience. I spoke of slaty-grey ones and found that he used to wear them years ago. 'Smocks are nearly gone out,' he said, 'but at one time all us shepherds wore them and when we went to Lewes Fair on the 21st September we used to make it Clothing Day as well as Fair Day. We always went to Browne and Crosskey's shop for smocks and gaiters and long gaiters and big overcoats and anything we wanted.'

A visit to Mr. E.A. Wheatley, the proprietor of Browne and Crosskey, of High Street, Lewes, confirmed these facts. Mr. Wheatley has been connected with the business for nearly forty years, and his father entered the firm in 1860. I was pleased to find that photographs of some of the old shepherds, which I was carrying, were at once recognized as those of customers.

Mr. Wheatley produced various items for my inspection: first some old documents relating to the firm, which although not connected with shepherds, were very interesting; then patterns of the drab and slate-grey linen of which Sussex round-frocks were made. A finished smock of each colour was shown to me; also a roll of the drab linen used for the purpose. The smocks were of the usual type, with excellent 'smocking,' but without elaborate ornament.

Orders for smocks are still received occasionally, but not from shepherds. The last sold were for the use of beaters at shooting parties. Our chat about the subject reminded Mr. Wheatley of a quaint incident which provided a new name for them. A countryman once entered the store and said: 'I want a 'og stopper!' He explained that when working among

— DOWNLAND SHEEP CROOKS — AND THEIR MAKERS

Downland Sheep Crooks

The Downland shepherd's 'badge of office' is his crook, which is used for catching sheep by the hind leg, or small lambs by the neck.

To one who has never seen a crook in use the first practical demonstration is very interesting, for shepherds become very skilful in handling these cherished tools.

Among the many inaccurate printed references to crooks is a paragraph published in a Sussex newspaper as late as April, 1926, which states that the metal crooks are the ornamental handles for the staffs! Yet it is easy to understand how such an idea originated, for a shepherd often uses his crook as a staff. For hours it might as well be nothing more, but when a sudden emergency arises its true value is apparent. It may be necessary to catch a sheep which has got into the wrong pen, or mixed with the wrong flock in sorting, but apart from these more or less regular happenings there are other causes for catching them. The quick eyes of the shepherd often detect need for doctoring, while inspection is frequently necessary in summer, when flies are a pest. Long brambles occasionally attach themselves to the sheep's woolly coat and must be removed, or perhaps a bell is to be adjusted. A dozen little happenings bring the crook into use every day, but so deft are the hands that wield it that once the shepherd has approached near enough the operation is over before you realize it, and the crook becomes a staff once more.

The long thin end of the crook, which is bent at an angle to the barrel is called the 'guide' and usually terminates in a curl. Thus a sharp point is avoided, and as the guide slips round the leg of a sheep no harm is done. The crook being raised as it catches hold, the leg of the animal is lifted and held securely. This allows the shepherd to handle the sheep and do what is necessary.

It must not be taken for granted that all the crooks used in Sussex were made here, for there are many shepherds about who brought their crooks and bells with them when they settled in the county; consequently, while preference must be given to the products of the Sussex forges, certain other 'foreign' crooks of elegant design also deserve appreciation,

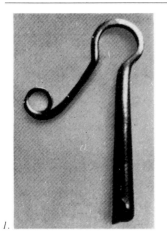

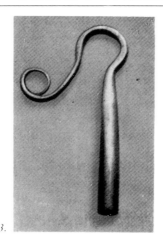

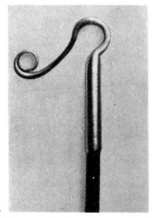

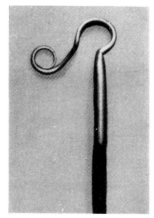

Sheep crooks: 'Nearly every shepherd had his own favourite.' 1. A crook of the famous old pattern made by Berry of Pyecombe. 2. The first brass crook made in Sussex. 3. A crook made by Green of Falmer. 4. A crook made by Hoather of Kingston-by-Lewes

for many of their owners have fallen under the spell of Downland and their crooks will probably end their days here.

What a wonderful collection of pictures of sheep crooks could be made in Sussex! Nearly every shepherd has his own favourite. It may be a gift or an heirloom, or it may have been made specially to suit his own fancy; whichever it is, there is generally some subtle feeling behind the matter – that feeling that makes one give preference to a particular tool, stick, rod or gun, and if a crook made to order does not quite satisfy him for any reason he will change it for another at the first opportunity. Some smiths have a wide reputation. The crooks made at Pyecombe were considered models in the old days, while those made by Green of Falmer and Hoather of Kingston-by-Lewes are of excellent design and workmanship, and are treasured by many shepherds.

I have acquired a few old crooks, with available data regarding them. As samples of one of the typical Sussex crafts they have a charm all their own.

Many of the very old crooks are fashioned out of wrought iron, while others are made from old gun barrels, with excellent results. I have heard of brass crooks, once made at Brighton, but have not seen one.[1] They were not very popular, and I have been told that many of them broke asunder and were thrown away.

In some cases the initials of the shepherd are engraved or inlaid in brass on the crook barrel.

In Downland it is rather unusual to find a crook in everyday use which is kept bright. The majority of them are allowed to be dulled by slight rust, although just a little effort would make them shine. Some shepherds have a 'best crook' at home which is bright, but it is kept for special occasions, such as sheep fairs.

Crook-sticks

The handle for the crook, known as the 'crook-stick,' is an important item. A good, strong, straight one is preferred. Hazel seems to be the favourite in Sussex, but ground-ash is often used, and occasionally a straight holly, cherry, or other stick is seen. Most shepherds endeavour to store up and season a few spare sticks. I have seen a live ground ash[2] which had been bent down flat and tightly pegged, when young, by a shepherd for future use. He showed it to me and cut it off while I waited with him.

Worn and weakened crook-sticks are replaced with new ones if possible, for if a stick snaps off short owing to age it is necessary to put the barrel of the crook in a fire to burn out the stick end. This is the reason why many old iron crooks are much worn in the barrel or have had that part shortened or repaired. A shepherd who treasures his crook would rather replace the stick occasionally than fire the barrel. To get a crook off its handle it is struck smartly inside the curved head with a heavy rod, or a poker, or anything suitable, and it drops off suddenly. To fit a new stick in the barrel is not so easy as one would think. To do so correctly the stick is carefully whittled to fit the barrel and put in lightly. The shepherd then tests it by resting the stick across his half-closed hand while his arm hangs by his side. The guide of the crook should point upward. If it topples to one side, it must be fitted on the stick again, otherwise the tool would feel top-heavy or difficult to manage. Directly the stick rides perfectly on the hand one is aware of the change, and a few smart taps on the ground with the end of the stick secures the crook in its place.

Certain vague references in print suggest that at one time it was customary in some parts for shepherds to sit and carve their crook-sticks, but in spite of careful inquiry I have been unable to trace any actual instance among Sussex shepherds. If it ever was customary in the county and the custom died out, the absence of evidence is probably accounted for by the comparatively short life of a crook-stick, whereas old wooden bell-yokes, made by shepherds, and often referred to as *wooden crooks,* were carefully finished and sometimes carved with numbers, initials, dates, and designs.

There is only one way to appreciate the many little details relating to the designs and uses of crooks, and that is to learn them from the actual makers and users; not from one man, but from many, and to handle such specimens of smiths' work.

My Favourite Sussex Crook

The crook was made at Kingston, near Lewes, about 1903 by Philip Hoather for Nelson Coppard, and was fashioned from a very good light gun barrel. Up to this time the shepherd had used a very old crook which had belonged to his father, and which was passed on to another shepherd when the present one was made. He has reason to remember the pattern of it, because it was one of his duties, as a

Through the thin light of the shepherd's hut, a special crook and favourite bells

'The charm of a hut is that one never knows what may be hiding in it – an old blue umbrella, perhaps, a bell or two, or an old crook. A small curio, a "shepherd's crown," or fossil, may rest on a ledge, a "dirt-knocker," or other interesting thing may hide among the bits of wood, wire, and other trifles that always seem to find their way into the hut. Occasionally a bunch of rabbit wires on a nail or a well-used gun tell their own tale.'

boy, to keep it clean and bright. The thin tail in the curl of the guide had worn very sharp and often 'ketched' his finger.

There is a distinct beauty in the present crook. Unlike so many of the ancient ones of wrought iron it has some elusive, indefinite charm when handled. It is as if the reliability of the barrel, which once made the gun a treasured possession, has been subtly transferred to the crook. The constant caresses of the shepherd's hard palm, the occasional rub on the legs of sheep, and the thousand touches of everyday use for twenty years have combined to wear away all the original sharpness of edge, and have given the metal a smooth, soft polish that artificial aid could scarcely imitate. The metal is very hard and is not easily marked; consequently each of the few noticeable dents is really a kind of shorthand record to the owner, recalling the various events which brought about their existence.

A dozen or more of his favourite ground ash sticks have done duty in the socket. They have all been about four feet in length from the end to the barrel of the crook. The shepherd,[3] unlike some, prefers a fairly short stick. He uses it as a staff, and likes to feel the smooth barrel in his palm when he is 'doddling round.' The present (and last) stick is marked at the lower end with several notches, like those on a tally stick, and it has practically served that purpose, as each notch represents a length of wire necessary for some fencing for a large fold. Another notch measures the length of an adder which was killed. The stick has been well used and had acquired a nice polish before the crook was given to me.

It was the first crook I ever saw in use. For that reason, and because I have heard the interesting tales which explain how the various dents and marks were acquired, and have so often felt the friendly clasp of the hand that carried it from the day it was made, my treasured keepsake is worth more to me than any other.

I felt that I should like to visit the spot where the crook was made, and I set out for Kingston Lane,[4] which branches off the Falmer–Lewes road. The first part of the lane, as seen on a hot July afternoon, was not inviting, for the road was dusty, the hedge-banks trimmed to bareness, and the tall hedges above cut back. Suddenly, however, the lane changed to a tunnel of foliage, a place of beauty and grateful shade – a most attractive entrance to the little village.

Mr. Hoather's cottage is in a little meadow. A large gate bars the entrance from the road, and another leads out of the meadow at the other end to a lane where the forge stands. I climbed the gate and found Mr. Hoather at home, for he retired from work on reaching the age of eighty. The sight of some shepherds' portraits roused his interests afresh, and he kindly accompanied me to the forge, where his son[5] now works. The original forge where he made Nelson Coppard's wonderful crook is by the side of the present shop, and Mr. Hoather stood by the door of it for a photograph, after which he adjourned to the other shop.

For fifty-seven years he worked at this spot, at the foot of steep ploughed land in the heart of Downland. Here came the oxen in the old days to have their 'cues' renewed and their brass horn-knobs fixed. Here came many shepherds to order their crooks and to have their bells repaired. Iron bells of the canister type were made occasionally, but not in any quantity.

Our mutual interests increased as we talked, and a search of the shop on my behalf resulted in the discovery of an old Pyecombe crook minus the barrel, a few of the last ox shoes made (which had never been used), and a knobbing iron – a

Mr. Philip Hoather, the blacksmith of Kingston-by-Lewes, outside the forge where he worked for fifty-seven years. He made Barclay Wills' favourite sheep crook for Nelson Coppard in 1903, 'fashioned from a very good light gun barrel'

tool specially made for putting brass knobs on cattle horns. These were gratefully accepted as souvenirs of my visit. Mr. Hoather's son is now in charge of the business, and is one of the few people who can and will repair a bell or crook to the owner's complete satisfaction. Like his father, he has that sympathy with the shepherds' requirements which results in carefully finished work.

Mr. Hoather accompanied me back to his cottage, and if I am able to climb a high gate as nimbly as he when I am eighty-one years old I shall be satisfied.

A Visit to Falmer Forge

By the Brighton–Lewes road, nearly opposite the Swan Inn at Falmer, stands a smithy: a very ordinary place to most of those who go to it, but of particular interest in its association with the shoeing of oxen and the making of crooks. The tools connected with ox-shoeing which are in Brighton Museum came from this smithy, and here, in the past, lived one of the Sussex sheep-crook makers named Green.

The unexpected possession of another of the famous Green's crooks brought a wish to see the forge again, and a chat with Mr. Slarks, the smith, who once worked with Green, revealed some items of real interest to those who own specimens of this craftsman's work.

Green gained a reputation among shepherds for good workmanship, and a feature of each crook made by him is the beautiful curl in the guide. From Mr. Slarks I learned that this curl was obtained in a particular way, for instead of being fashioned on the point of the anvil the thin tail was curled into a circle on a curious little home-made iron contrivance. It is about four inches high, and was fixed in the vice in front of the forge. The slate carrying a scratched outline of a crook, which Green used as a pattern, cannot be found now, neither can the tool with which the deep groove was made in the crook head, but this curious iron curler is an interesting link with old days.

I came away with the impression that Mr. Slarks holds less reverence than I do for such links with the days of crook making at Falmer. If I had the same recollections as he has perhaps I should not be so enthusiastic.

Green was in charge of the business, which was owned by Mr. Carter, consequently he was able to give special attention to certain things in which he was particularly interested. The making of a crook was a delight to him, but if the truth be told it afforded little pleasure to his assistants. At such times his anxiety for perfection in every detail made the atmosphere of the smithy very unrestful – his rapid instructions and admonitions are yet remembered – and his pleasure with the finished work could not make up to his assistants for the tense moments round the forge. Most of us remember some distasteful tasks done in youthful days, and it is easy to understand that Mr. Slarks has never found any real enjoyment in crook making since that time.

It is probable that if all the hours spent on these crooks had been properly charged up to the job it would have been found that the cost of making was far more than the small sum received for the article. It is therefore no wonder that many shepherds availed themselves of an opportunity to secure a crook which was really a bargain at the price.

Pyecombe Hooks

To anybody interested in shepherds' crooks there is some magic in the words 'Pyecombe hook.' Some shepherds certainly refer to their crooks as 'hooks,' and the term is often used for the Pyecombe product, although the reason is not apparent. However, as Sussexians say, 'there 'tis!'

Crooks made by a smith named Berry at Pyecombe forge gained a wonderful reputation among shepherds of Downland, whose portraits now adorn the homes of their descendants. An authentic specimen is regarded as a curio, but it is seldom that much importance can be attached to the label on a crook simply marked 'Pyecombe.'

During my rambles I have found two of these ancient crooks. The first was in the possession of Mr. Walter Upton, shepherd at Brown's Farm, West Blatchington, and was last used by him about 1916. It was made by Berry for his father. There is no record of the actual date, but as it was before he first saw the light it must have been previous to 1864.

The second 'Pyecombe hook' is a good example of the particular pattern which was so much appreciated. It has the letter B stamped in the end of the groove where the bend of the guide commences. It was last used by Mr. R. Flint of Findon in 1924, and was passed on to him by another shepherd at Findon Fair. Its barrel is nearly worn away, but the other part is still in fair condition. The details of this type of crook were described exactly to me before I became the happy possessor of Mr. Flint's specimen.

After finding these two crooks I felt that I should like to

Pyecombe forge, where the ancient crooks, with their 'wonderful reputation among shepherds of Downland', were made by Berry

for eleven years, but had forgotten other names which would make the total at least seventy years. The latest work by Berry therefore appears to date from about 1855. One of the tenants enlarged Berry's workshop and built a bigger forge, which is the one in use, but Berry's original forge is still there in the shop and can be seen by those who care to make a pilgrimage to the spot where the famous crooks were fashioned.

Mr. Mitchell is very active, although eighty-one years old, and still makes a crook occasionally. He told me about the oxen he used to shoe and many other interesting things. On his advice I went to find the one shepherd of the old school left in the district, Mr. Wooler.

I took a roundabout path, as one often does in a fresh locality. A small stack offered shade from the scorching sun for a while, and while resting I was amused by the antics of swarms of mice, which peeped out from the crevices all over the stack. I shared my biscuits with them and then worked round Wolstonbury Hill by a lane bordered with speedwell blossoms and a track through the glorious high June grasses. As I reached the brow a few sheep moved away, and I heard the familiar call of canister bells.

I soon found Mr. Wooler on the shady side of the hill, with two dogs. He wore two bee orchids in his hat band. He was brought up among sheep, as his father was a shepherd, and the canister bells were handed down to him. Some of them are much worn and some have been mended. Mr. Wooler confirmed the fact that the tone of many bells is spoiled when repaired, but he spoke warmly of old Mr. Mitchell, because the smith's sympathy with, and knowledge of, a shepherd's requirements, in addition to his skill, make a great difference to the work he undertakes.

Mr. Wooler agreed that the old shepherds are dying out. In view of the fact that times have changed so much he

see Berry's workshop where they were made. I travelled to Pyecombe on a June morning (1925) and found the forge on a hill by the side of the church. The business has been carried on for the last fifty-four years by Mr. Chas. Mitchell, who explained that several other tenants preceded him between his entry and the time of Berry. He could account cannot honestly recommend any young man to adopt his profession, as it will soon become an unthankful job for anybody. Once more I heard, as I had done from Mr. Mitchell, regret expressed by one of the older generation that 'shiremen' have acquired so many Sussex farms.

Mr. Wooler's crook was made at the Pyecombe forge about fifty years ago. It is now beginning to split in the bend, which means that its term of years is nearly run, as repair, to his idea, would not be satisfactory. It is one of his treasured possessions and an example of Mr. Mitchell's early craftsmanship. Its fifty years of wear have certainly earned for it the right to rank with the others and have invested it with some of that subtle attraction and renown accorded to all old 'Pyecombe hooks.'

Mr. Aucock, another shepherd, who was born at Alfriston, lived with Mr. Funnell, and on the occasion of our visit an impromptu 'bell inspection' was hastily arranged for our benefit. First came Mr. Funnell's brass bells and a big horse bell. Then, to our delight, Mr. Aucock opened a bag and tipped out a collection of canisters, Lewes bells, and tackle. It was a wonderful experience to find two shepherds, living together, with collections of the two most uncommon types of brass and iron bells in their possession. I like my photograph of the scene which greeted us on 'bell inspection day' – a simple, homely picture, but a reminder of that kindness so often shown by Sussex shepherds and their families to those enthusiasts who take a sincere interest in the details of the shepherd's life.

CUP BELLS

'Cup' bells are aptly named, for they are like little inverted cups or bowls. They are fitted with tops or staples of various designs for attaching them to the tackle. They were made in sets, and being cast from bell metal, were made in tune one with another. A set must have made a merry jingle! One old shepherd, referring to his father's set, said: 'They was a band of music, an' no mistake!' Another described the notes of the small cups as 'dingle-ding, dingle-ding!'

There do not appear to be any complete sets of old cup bells now, but even a few, ringing together, produce a very pleasing effect.

Although I have some which are quite plain, cup bells are generally ornamented with incised lines round them, which vary in number. It is unusual to find a specimen with its original crown ring and tongue, possibly because those parts were rather thin and light.

HORSE BELLS OR 'LATTEN BELLS'

The gradual disappearance of horse teams wearing frames of bells created many opportunities for shepherds to add to their collections. Beautiful old frames have been broken up so that the bells could be mounted for use on sheep, and many lots of clucks and dull-sounding bells were made more cheerful by the addition of this type of bell. Some shepherds were tempted to acquire a lot, and one man, who has now left the county, succeeded in collecting a large number and used them all on one flock. The result was a deafening jangle. The old-fashioned charm of chimes from eighteen bells in frames as the team drew a load was completely lost amid such a din.

Many of these bells are cracked by striking stones and troughs and thus rendered useless. Through this rough treatment their number is gradually diminishing – the sad result of using beautiful bells which were not intended for such a purpose.

The original crown rings and tongues of these team bells were so thin and dainty in design that most of them have worn through and have been replaced with others of stouter make. The initials R.W.,[4] or the name R. WELLS, in raised letters, found on the inside of many horse-bells, refer to the maker.

WHITE LATTEN BELLS

Among the 'horse' bells used on sheep are sometimes found a few of another type known to shepherds as 'white latten bells.' These are usually numbered on the crown, showing that they were made in sets, as the horse-bells were. Very little is known about them, but shepherds are eager to retain those they have, as they are very pure in tone, and sound well in the open, especially when mounted on chin-boards.

In appearance these bells approach most nearly to old house-bells, but they are of good silvery metal. Their crowns are somewhat flatter than those of horse-bells, and their rims are thickened with a band, and sometimes ornamented with circular lines.

The note as to tongues and crown rings of horse-bells applies equally to this allied type and specimens with original thin inside fittings are rare.

CROTALS OR RUMBLERS

I have known only three instances of the use of these bells on sheep in Sussex.

The crotal is an ancient type of bell which has survived until the present day. Small specimens are still found occasionally on horse harness. I have been told that they were mostly used as cattle-bells, and I own a large one (4½ inches in diameter) found on Newmarket Hill near Falmer, which still has an iron fitting attached to hold a broad strap such as is used for a large cow-bell.

'Rumbler' seems a good name for a large crotal, for the note is peculiar and is caused by the continuous rolling of a small iron ball inside. In old specimens this movement has not only worn away the edges of the slit under the bell, but has also caused the iron ball to wear into an irregular shape, consequently much less sound is given than when the bells were new. A glance at the slit, the iron ball, and the hole in the flat appendage to the crown will show whether a specimen is old or modern.

The initials R.W. incised on these bells (which are also found inside many team bells) stand for Robert Wells, the bell maker of Aldbourne, Wiltshire (1764 to 1825). Messrs. Mears & Stainbank, of the Whitechapel Bell Foundry, London, who took over the foundry at Aldbourne in 1825, have the original patterns of these 'sleigh bells' and use them at the present time.

Shepherds who have few bells and wish in vain for more, do not despise the ordinary old-fashioned house-bell, nor do they throw away a broken bell if it can be utilised in any way. Strange specimens are found. One used by Mr. Newell on Dyke Hill was half of a large crotal or 'rumbler,' fitted with a crown ring and tongue. Another which Mr. Nelson Coppard had among an odd lot he purchased was strange to both of us. Eventually we discovered that it was the top part of a large horse-bell fitted with a little tongue.

I have found the study of sheep-bells a fascinating one, for apart from interest in actual specimens the collector gradually understands and shares that deep regard which the shepherds have for their simple treasures.

Of all the bell music heard on the hills the songs of canisters are most charming. More tuneful than clucks or other iron bells, yet not severely correct in tone like cast bells, the canister's mellow notes have a strangely capti-vating sound. To hear the chimes as their wearers go eagerly homewards across the Downs is to capture a fadeless impression which breeds a desire to enjoy such pleasure again.

Tackle for Sheep-bells

A glance at the illustration of bells shows that several kinds of 'tackle' are used for fastening them on the sheep's necks, and although quaint little fittings are occasionally found, the tackle in general use is made to more or less standard designs.

At first sight this bell tackle may appear an unimportant item, but the fact that the various parts were always made by the shepherds and could not be purchased at shops, is enough to create interest in such specimens of hand-work.

All the usual items in a shepherd's gear are necessities, but his bells are an optional addition; and although they have a definite use they often prove to be his one hobby.

In past days shepherds collected them and found such pleasure in their possession that they spent many hours of their short leisure in making carefully finished tackle for favourite specimens. It was no tiresome task, but 'a matter of personal pride and delight in the job' as one man expressed it. So careful and so thorough was the work in these yokes, pegs, and chin-boards, that some have lasted forty or fifty years. The original sharpness of outline and the marks of the knife have been nearly worn away by the daily rubbing by leather straps and greasy wool, until wood and bone have that smoothness and softness which only age and constant use can produce.

Care of the bells and the making of tackle for them has been a matter of such importance in the lives of many shepherds that I have made a few notes regarding the various methods adopted.

WOODEN YOKES

Sheep-bells were often suspended by straps from wooden yokes, made from pieces of branches of trees. Yew was the favourite wood on account of its toughness and lasting quality. The sharply bent branches of yew trees provided many arched pieces suitable for the purpose. Other kinds of wood were used when yew was not obtainable, also juniper and furze branches, but the softer kinds were not very satisfactory.

These wooden yokes, often referred to as 'wooden crooks,' passed through several stages before they were completed. After being kept and seasoned in the rough state they were split or chopped into as flat a shape as possible. Sometimes a thick piece of branch could be split in half and made into two crooks. Holes or slots for straps to pass through were made, sometimes by cutting, sometimes by boring or burning, and were finished off with a knife. The careful knife-work then commenced. A hard new yoke took some time to finish off, chip by chip. Occasionally the shepherd's initials were carved on the 'crook,' and in some cases numbers (probably referring to the number of the bell in a set) are found. Less frequent are carvings of dates and devices. I have three yew crooks which are marked: (1) H.K. 1863; (2) 1887 with a star on either side and (3) 1888 with a flag and a cross.

A coat of paint in farm-cart colours, bright red or blue, was the finishing touch. Many old yokes still show some patches of red or blue paint, or traces of both, for at one time it was customary to renew this once a year when bells and tackle were overhauled. Yokes were then threaded on a wire and hung up before being painted. Referring to this a man said to me: 'I don't know how 'twas, but there seemed to be time for everything those days! What with old 'uns an'

'It is a pleasant sight to watch the arrival of a thirsty flock. In a few seconds the peaceful solitude of the pond is transformed. Its edge is thronged with eager animals, and the noses of the first few disturb the surface of the mirror. Without a pause the crowd encircles the pond. The old bells on the ewes have their edges immersed while their owners take the welcome drink. After much pushing the noisy crowd are satisfied. Slowly they move away . . . only the footprints and the wetted edge tell of the visit of the flock.'

Dewpond by Chanctonbury Ring

noo 'uns, I always had a tidy row of 'em on the wire. Some was blue an' some was red, an' once I painted some in stripes, blue *and* red, but I altered 'em nex' time!'

LEATHER YOKES

Inability to acquire suitable branches for 'crooks' led many shepherds to adopt an easier plan. Thick pieces of harness leather were trimmed to shape and slots cut without much effort. Such yokes are common, being easy to renew at any time, but they have no charm like the old wooden ones. They are not painted. They are frequently referred to as 'collars,' and the term 'crook' is reserved for the yoke made from a bent branch.

STRAP COLLARS

A strap collar, usually known as a 'strap,' is the simplest form of bell tackle. It is just a stout buckle strap with a little hole cut out to receive the top of a bell. Horse and latten bells, crotals and one kind of cup bell, all of which have a little flat appendage with a hole in it on the top of the crown, are easily mounted in this way. The bell-top is pushed through the hole in the strap and secured with a split-pin.

Bells with two staples are occasionally mounted by putting the staples through a slit in a buckle strap and securing them with pegs. In course of time the staples wear deep grooves in the wooden pegs.

Buckle straps are used for most bells with one staple set from front to back. If necessary wedges of leather or a wooden peg, or both, keep the bell in position.

CHIN BOARDS

The rough-and-ready method of fixing certain bells to a strap collar with a split-pin did not satisfy the shepherds who made their own wooden yokes and pegs. They cut out 'chin-boards,' which were small pieces of strong wood (generally oak) about six inches by three. A small hole in the centre was shaped to take the flat pierced appendage on the bell-head and a split-pin above the board held the bell tightly. A leather or iron washer was used under the pin if required.

Two varieties of chin board are known to me. The first has two slits cut through it (one each end), and is used on a stout buckle strap which passes through and under it, between it and the bell. The second variety has four holes (two each end), and straps are threaded through before being hung from a yoke.

When I acquired my first chin-board a shepherd told me that in his young days it was customary to use pieces of oak palings for making them. 'They was good palin's then,' he said, 'made o' good oak. There was mos' times a few bits broked off, lyin' 'bout, if so be you wanted some.' Then he remarked drily: 'Nobody seemed to know how 'twas the tops o' palin's snapped off, an' nobody looked for the bits under the sheepses' necks!'

I have one chin board, made by Mr. Newell of Devil's Dyke Hill, which is cut from a thick piece of leather instead of oak.

Bells mounted on chin boards not only stay in position, but give sharper, clearer notes when jerked, as the sheep are feeding, than those which dangle from straps.

BELL STRAPS

Straps used for hanging bells from yokes call for passing comment. They vary in width and thickness, according to the sizes of the staples and slots through which they pass. Sometimes a second (but much shorter) pair are put inside them for greater strength and rigidity. The four thicknesses of leather are then bound round with string or fine wire. Occasionally longer straps are used, being threaded through the bell staples twice instead of once.

REEDERS

Reeders are rough leather 'washers' sometimes used between a yoke and the peg to make a bell hang level. Any difference between the length of the two straps or the curve of the two ends of a yoke is thus corrected, and the bell hangs comfortably on the neck of the sheep.

LOCKYERS

Lockyers are the pegs which hold the straps to collars and yokes. They are mostly of wood, sometimes of bone, and occasionally of leather. The latter are merely makeshifts, but those of wood and bone claim a share of the interest given to old wooden yokes, for the oldest ones, deeply grooved with wear, and polished by years of use, have a charm of their own.

As in the case of yokes, yew was preferred for lockyers, and some specimens have lasted many years. They were cut out with a knife by shepherds. The rougher modern pegs are poor things by comparison.

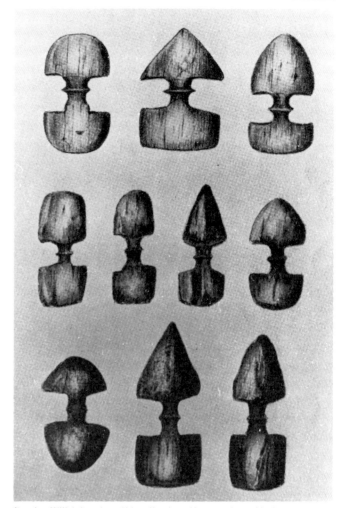

Barclay Wills' drawing of his collection of bone and wood lockyers. Made of bone, wood – preferably yew – or occasionally leather, they were used to secure sheep bell straps

Bone lockyers were cut from rib bones with the aid of a vice and saw. Such work was only undertaken by those who took pride in their bells and tackle, and they were pleased to secure good bones for the purpose. Quality of bone and workmanship vary in different specimens. I have some so beautifully made that after many years' daily use they are as smooth and pleasant to handle as ivory.

The two deep grooves in the 'waists' of lockyers are cut out very gradually by the movement of the straps round them.

Some of the set of brass bells at Wilmington already mentioned were fitted with long *single* straps, twisted and pegged through strap collars with bone pegs, consequently all these pegs have only a single groove cut in them and are an interesting variety.

All old lockyers are worn with a slight hollow on the under side which has rested and rubbed on the leather or wooden yoke. As they were designed according to the material used, and to the fancy of the maker, many varieties of shape are found, especially in those of yew wood. Although each one was made as an essential part of certain bell tackle, a collection of them is now of great interest, simply as an example of the careful handwork of the oldest shepherds.

My interview with Mr. Blann records his remarks on the making of yokes and lockyers. The quaint little saw is over a hundred years old. Mr. Blann inherited it from his grandfather, who was a wheelwright at Sompting. The work was done mostly in the evening, but a little of it could sometimes be finished off while on the hills with the sheep.

If any excuse be needed for such a long chapter on the details of old sheep-bells and tackle, the record is surely justified by the fact that these things played such an important part in the lives of many shepherds. Bells were useful; bells were company; bells provided a good reason for a hobby and an outlet for industry and keen interest which sweetened the hours of duty. The planning of well-finished home-made tackle for them brought into the shepherds' lives the secret satisfaction of the collector, and the pride of the craftsman in work done for pleasure, and for this reason bells and tackle deserve their due share of space in any record of the lives of the men to whom they meant so much. The old shepherds have so little to leave behind them that these carved pegs, yokes, and chin boards will be treasured one day as specimens of their work.

A Ring of Canisters

We are quite used to references to 'rings' or 'sets' of bells of various kinds. We are familiar with the rings of horse bells which were hung in hoods, and we know that other sets of cast bells are still made, but in the case of the ancient 'canister' sheep bells there is more to record.

Although I had often heard that a ring of canisters consisted of eighteen, or twenty-one, or sometimes twenty-five bells, I had no definite knowledge of the details. I often wondered about this: was a real ring a run of notes as on the keys of a piano, or was it a series of chimes made up in some particular order? I guessed at the answer, but I could not be sure that I was correct. I wanted definite proof instead of the rather vague information given me. Such sets as I found proved, on close inquiry, to be composed of old rings that were incomplete or had been made up with extra bells, but now, after studying a real ring which was loaned to me, I am able to follow the making of a ring and to piece together the scraps of information which I have gathered at various times.

The note of the largest and lowest-toned bell in this ring is A, which starts a chromatic scale of twenty-one notes.

The suggestion that only approximate notes could sometimes be obtained is proved by these bells. Although the principal chords chime nicely, many of the notes between are not quite true.

Finding some difficulty in arranging them in correct order, I hung them up in two lots on sheep-crook handles, one lot starting on A, and the other on A sharp; and the fact that some bells are duplicated causes me to think that I have by chance discovered proof that a ring was made up in this way for convenience. As one example – the eighth bell in the lot starting on A and the seventh in the lot starting on A sharp are duplicates. This accounts for two bells of one note being included. There are three duplicates in the ring.

Further proof is given by the extra bells, pointed out by the owner as tenor bells. They blend with more than one bell in the set to which they belong. These tenors were extras to each lot as I had them arranged. I already own four of these tenor bells which ring a double note. They could not be included with satisfactory effect in a scale, and the actual proof of similar bells as additions to a ring was very interesting.

One of the small bells appears to be missing, although, in the open, this would not be a noticeable defect. Possibly the owner was unable to obtain one of the required note.

The smallest bell, which has the highest note, is only three inches high by two and a quarter wide, and by all the evidence I have such tiny ones must have been far more uncommon than the larger sizes. Probably there was less demand for them. Some old shepherds now living do not care for tiny bells – they prefer the songs of the good medium-sized ones, accompanied by the deep tones of one or two of large size, and not every shepherd would trouble to acquire a real ring.

It was evidently the intention of the owner to acquire a bell to fill the place of the missing one if opportunity offered, for with it the set would be complete. Several additional bells proved to be duplicates, of good power, and were probably kept with the ring for this reason. Their presence could not interfere in any way with the rest.

The largest bell is so large and deep-toned – it is over six inches high – and the smallest, as already stated, so tiny, that it would probably have been unusual, even at the time when rings were collected, to make up twenty-five bells in one run of notes, but four tenor bells (one medium-sized and one small to each of the two scales) would increase the ring to twenty-five. The only other arrangement would be the use of unlimited duplicates.

It cannot be stated that a big bell which sounded the note of A was always picked to start a ring. A run of eleven bells starting from another note, combined with ten starting from the following one, would still make a ring of twenty-one, and so on; but it must be remembered that a start on a higher note than A would make it necessary to include in the ring more of the tiny bells of still higher tone than the smallest used here.

This set may be taken as a good typical example. I proved it by arranging some of my own bells in similar order. I made a complete ring, but to do so I was obliged to include a few horse bells for the highest notes. Even with my collection spread out in a room my ring 'wanted some getting together', as a shepherd had assured me. I can imagine that the selection of a set from a pile in a crowded, noisy fair-ground must have been somewhat difficult, and can readily understand that perfect tuneful notes could not always be obtained unless the ring was made up by the actual makers from their stock.

From my investigations it appears that a ring of canisters is really eighteen bells, sometimes increased by three duplicates or tenors to twenty-one bells, or by three duplicates and four tenors to twenty-five bells.

It is so seldom that a real ring of old canister bells, proved by its history, is available for reference that it seems quite worth while to record the full details of this set, as furnished to me.

The bells were bought, when new, at Lewes Fair, by shepherd Harry Strong of Cold Dean Farm, near Falmer. To his friend Solomon Gorringe, another shepherd, he once said, 'When I have done with them, Solly, they shall come to you, or to one of your boys.' The transfer happened sooner than either of them expected, for Strong was badly injured at Lewes Fair when a steam-engine exploded, and after his death Mr. Gorringe had his friend's bells, as arranged.

The new owner treasured them and guarded them jealously, and was very distressed when he lost one of the precious ring on the downs. Upon hearing one which he thought was like it he claimed it, but it proved to be the property of the shepherd who had it, and was marked with his initials, J.E. Mr. Gorringe bought it to replace the missing bell (although I found that when in its place in the ring it is not a very good substitute for the lost one). Mr. Gorringe's son Albert (also a shepherd), who inherited the ring, knows that this is the only alteration made to the ring in his father's time.

First used on Hollingbury Hill, and subsequently on the hills at Plumpton, Pangdean, and Devil's Dyke, this ring for

Albert Gorringe of West Blatchington. It was through his kindness that Barclay Wills was, after much searching, able to find an old ring of canister sheep bells

many years made music over the stretch of Downland behind Brighton, but since the death of Mr. Solomon Gorringe, on December 30th 1907 the bells have not been used. Whether they ever will be remains to be seen, but at present, although they are still lying idle, they are not for sale at any price.

It is always pleasant to find what one hunts for, particularly if the chase has been a long one. After four years' study of the rather unusual subject of sheep bells, I am deeply indebted to the West Blatchington shepherd, Mr. Albert Gorringe, for enabling me to print the actual details of this canister ring. In addition to being a record of one of the small and simple things which might have faded out without any note having been made, the information gives a very necessary finish to any serious account of the old Sussex sheep bells, which were once such a feature of the county.

— THE SHEPHERD'S —
COMPANION

So many of us are dog lovers, and fond of our jolly companions, that we hardly realize how different is the lot of the sheep-dog[1] to that of the ordinary pet. He (or she) is born to a life of responsible work, and shares with the shepherd the care of the flock. Like him, he works every day in the year, and his leisure is very short. The walk towards home at the end of the day is generally his only opportunity for play, and then, if he is not too tired, he may poke about and hunt, or do any of the hundred things which delight him. Occasionally a shepherd, whose own heart beats quicker at the swish of a bird's wings or the sound of some animal in the undergrowth, will allow his four-footed mate a little liberty if occasion offers, but the word *duty* looms so large in front of them both that the shepherd cannot often afford to give the signal for relaxation. More often he finds it necessary to shout: 'Come back, you monkey!' or something equally curt, although there are times when such expressions are not a true index to his real feelings.

On meeting old M—— you might think him very gruff, and to hear him shout at his dog you would be tempted to form a rather unfavourable opinion of him, yet you would be mistaken. 'I do bawl at that dog,' he remarked, 'but he do need it, times – I don't know whether I ever *shall* make a good dog of him, but I'm afraid not! It don't do to let him

see me laughin', but he doos some comical things. Times I bawls at him, but I laughs inside me all t' time. I reckon I be gettin' reg'lar soft – but I bin a nursemaid ev'ry spring fur nigh on seventy year, so p'raps that be somethin' t' do with it!'

A shepherd is obliged to be particularly strict with a young dog in training, hence his impatience with people who feed a sheep-dog or bring other dogs near enough to distract the puppy's attention from the work in hand. It is usual to put a young dog to work with an older and experienced one for a time, but directly it shows an aptitude for work it should be separated from its companion and trained to obey orders alone. 'It is a great mistake to leave them together too long,' said one man to me, 'for if 'tis done the young un is apt to expect to wait for t'other to help, and not *try* to work alone.'

It is the careful, patient training of dogs which is responsible for the wonderful work which surprises us. Every shepherd has his own ways, and his dog is used to them, consequently we may see dogs obey signals made by a raised arm, or a motion with a crook, or respond to nods of the shepherd's head, movements of his eyes, whistles, calls, muttered words, and other half-secret signs.

The following incident was related to me by a friend who

had been studying the ways of shepherds in connection with a book which he was writing:

While there was a mist on the hills I went to find Shepherd G——, just for the experience. I wanted to hear unseen sheep bells and to study the details of a shepherd's work at such a time. After much difficulty I located the flock and G——'s collie found me. As you know, he is not a very safe animal. He rushed at me when I appeared, but fortunately did no harm. When I met G—— I told him that I had had difficulty in finding him and that I wished to accompany him in his walk through the mist. To my surprise he asked me to come another day, and on being pressed for an explanation said: 'You see, I want all the work I can get from the dog in this mist. If *you* come he'll be trying to watch you as well as the sheep, and I can't afford to have him attending to anything except his work.'

Although I was disappointed I could appreciate the shepherd's view of the matter and was glad to have heard his explanation, as it gave me a fresh idea about the work of a sheep-dog.

Some shepherds show a preference for bob-tails or similar rough-haired dogs, others for collies. Both kinds are seen at sheep sales, chained to pens or wheels or other convenient places.

The sheep-dog sees more of his master's ways than anybody. They start off together when the day is young, be the weather what it may. They are often companions in misfortune. Drenching rain, stinging hail, driving snow, or piercing wind may be their lot – it makes no difference. With bowed heads they plod along to do their duty among the waiting flock, and if we ever wake to a raging storm or a

John Beecher with his dogs

bitter winter gale, and think of some old hill shepherd and his dog, it is the steady, reliable, rough-haired bob-tail that seems the fit companion for the grizzled veteran with the crook.

On meeting one shepherd soon after shearing time I looked for his usual companion – a bob-tail with a very long shaggy coat, so matted together with mud that it would have defied any brush and comb. Instead of the mud-coated dog I saw a fresh one – a thin, gaunt creature, and I remarked: 'You have a new dog, shepherd – what has happened to Bill?' The shepherd smiled. ''Tis still Bill!' he said, 'but he wur in sucha pickle wi' mud in his coat that I sheared him! I'd got the shears handy, you see, an' I said to him: "I'm damned if I don't do you as well, Bill, now I be at it!" He beant hardly used to it yet, but he couldn't enjoy himself when the weather turned warm – still, come the winter, reckon he'll be the same ol' Bill again!'

Among many pleasant memories of sheep-dogs and their owners I recall a meeting with John Beecher, a shepherd lad, in a valley east of Cissbury Ring. John was devoted to his work, and content to do it regardless of what future years might bring to members of his profession. With him were two rough-haired dogs, an old one and a young one which he was training. There was no water supply on the feeding-ground, so every day John took a large bottle of water with him, which was far more than he required, and after lunch he turned his empty waterproof dinner bag upside down, punched it in to form a bowl, and filled it with water for his shaggy friends.

I asked him to hold my staff and pose for a photograph so that I could use it as an illustration for this little record of his thoughtful action.

Collies used for sheep-dogs are often unsafe for strangers to handle, but Prince was an exception to the rule. When I met Shepherd Newell near the Devil's Dyke he had a fine collie for a companion. Newell's flock was usually penned for the dinner hour in an enclosure near the Dyke Hotel, consequently Prince found many friends. It was only natural that constant attentions from visitors made him 'soft' from the shepherd's point of view, but his master, a dear, quiet old man, smiled indulgently when ladies exclaimed: 'Oh, what a beautiful dog!' and fed him with dainties of all kinds. 'He gets plenty of nice chocolates and things given to him,' said Newell, 'but the ladies don't think to say: "Will shepherd have one?"'

I stayed in the enclosure while the shepherd had dinner, which was brought to him from his home, and Prince sat very close beside me on a bank. I shall never forget that meal. The pieces I gave him seemed so small and inadequate, and were gobbled so quickly, that I was almost ashamed to look into those beautiful eyes and note their wistful, pleading expression. Prince soon finished my lunch; his long soft nose went to the bottom of the bag and not a crumb was wasted. Then I whispered in his ear, and apologized to him for the shortage, as we sat cuddled up with our heads together, and he kissed me to show that he understood what I said. Call me silly if you like – but any dog lover would have done the same!

— A Dewpond Chapter —

Much has been written about the Downland dewponds, or, as the shepherds prefer to term them, sheep-ponds, but in the face of all the conflicting opinions upon the subject it is very difficult to decide as to the source of their supplies. One very clever and well-known author[1] devotes twelve entertaining pages of his book to a record of the evidence he collected regarding them, but considers that no final opinion can yet be given. That being so, it is well to leave all discussions on the point to the authorities and to take what enjoyment we may from a survey of the ponds and their varied visitors.

It is a pleasant sight to watch the arrival of a thirsty flock. In a few seconds the peaceful solitude of the pond is transformed. Its edge is thronged with eager animals, and the noses of the first few disturb the surface of the mirror. Without a pause the crowd encircles the pond. The old bells on the ewes have their edges immersed while their owners take the welcome drink. After much pushing the noisy crowd are satisfied. Slowly they move away. The dog watches his master's face for the signal to refresh himself. It is given and he leaps to the pond edge and laps vigorously. His feet and legs are wetted and his face well washed. He somehow gives you the idea that he could enjoy much more if only he could spare the time. He stays on the bank just long enough to give his head and feet a quick shake, then stands at attention. The shepherd raises his arm. Instantly the wonderful dog obeys the instruction. The sheep have not had time to stray far and are soon rounded up and gently turned towards the track leading to their next feeding ground. The pond is again deserted and only the footsteps and the wetted edge tell of the visit of the flock. The surface is a mirror once more and will soon reflect the forms of other smaller visitors, for many birds come to drink and bathe. Dainty wagtails trip nimbly round the pond or fly across it in pursuit of insects. Linnets, finches, and other small birds come at intervals all day long. Rooks and pigeons are often seen, while occasionally a party of magpies may be observed. It would be necessary to have a hiding-place near the pond to learn the full extent of the visits of various creatures. My collection of odd feathers found on the few occasions when I have visited dewponds has included those of the tawny owl, thrush, pied wagtail, starling, rook, jay, magpie, cuckoo, partridge, and ringdove.

I once endeavoured to probe the mystery surrounding the dewpond theories, but the result was a dismal failure. I was lunching with a shepherd on a bit of high ground near a dewpond. His collie lay beside me and her beautiful eyes looked wistfully at my biscuits and then at my face, and then at her master. I asked permission to share my meal with her, and the shepherd consented, though he remarked, 'Reckon yew'd soon spile 'un!'

On the slope below us the sheep were feeding quietly.

Nelson Coppard and his flock at a dewpond near Pyecombe. The picture was taken while Nelson was shepherd at Pangdean Farm. INSET Barclay Wills' picture of his friend the shepherd, Nelson Coppard

— SHEEP IN DISTRESS —

The Shepherd's own Page

'Reckon you can spare one page out o' your book for me,' said my friend the shepherd.[1] 'I bean't much of a scholard, as you know, or I'd have writ to the paper long ago, for 'tis quite time folks was told about helping sheep up when they be rolled over.'

He had seen two men some distance away who stopped, looked at something, and passed on. Soon afterwards two others did the same, but came in his direction. Seeing my friend with his dog and crook, one of them left the path to tell him that they had seen a sheep lying upside down in a hollow. It could not get up and was 'as fat as a barrel.' 'An' yet you diddun' try to help 'un! What sort of a man be *you* then?' said my friend hastily, and without waiting for an answer he ran and set the sheep on its legs. 'But she wur near gone,' he told me. She belonged to a flock on the next farm.

The shepherd has saved many lives in this way; hence his request for a page on the subject.

Sometimes when a sheep rolls it cannot regain its footing, especially if it be fat. Soon the animal swells, and dies in misery if no help comes. There is no need to hesitate about touching a sheep – it will not bite you. It is the duty of anybody and everybody who uses the Downland paths to take the trouble to set such a sheep on its feet and to save its life, or at least to obtain assistance for it quickly.

Should any readers meet with a sheep in distress I hope they will remember the shepherd's appeal and give what help they can.

Under the Turf

In spite of the fact that one of the worst samples of weather had just been experienced in the Worthing district I took advantage of a bright spell on September 25th to make an early start, and by the time that the smell of fried bacon was being wafted about in the town I was treading a drenched field-path of sticky, slippery chalk; past swedes and scarlet poppies on one side and a low hedge of thorn, sloe, and bramble on the other. Finches and tits and butterflies were already busy, and a Red Admiral, poised on a tall blue bell-flower, left its perch and settled on my coat for a moment as I stopped to pick a few juicy blackberries before I turned the corner and ascended a steep track over the hill.

After passing the brow I found the company I expected – hawkweed and yarrow, scabious and harebell, some small ox-eye daisies and a few late rampions. Further on, where the path was almost lost in the longer herbage of a bottom, was a mass of burnet rose bushes, many of them bearing their black fruits.

Upon the next brow I could see a big chalky heap, which denoted a fresh excavation, and I went direct to the spot, hoping that I might be lucky, as I have often been, and that I might find a flint or other treasure left behind for me. The pressing attentions of three small blackberry-pickers, who saw me clean and pocket a little implement, drove me away, but I discovered another promising-looking place where a flint mine had been opened and filled in.

I was stepping aside to avoid a hole in the turf, about a yard in diameter, when a sudden slight movement caused me to stop and look down into the hole. I expected to find traces of rabbits or something similar, and was not prepared for the surprise which awaited me. In the cramped space below were four sheep, standing with lowered heads. Quickly I plucked a bundle of sweet, short grass blades, which I dropped on the back of a sheep whose head was quite hidden in the recess, and the nearest prisoner devoured it eagerly. My attempt to take a photo of them was a failure, for the hole was larger below than at the opening, and very dark. My next thought was to obtain help, for it was quite impossible to get into the hole and lift them out, but I had scarcely started to look round the hill when a sudden heavy shower compelled me to run for shelter to a

'Stumpy' Arnold, the Burpham rabbit catcher. He spent the winter in a hut on Blackpatch, existing mainly on a diet of pork, whiskey and tea. He began working at the age of seven and caught an average of four thousand rabbits a season

clump of trees and bushes, where I was obliged to remain longer than I wished.

At last I was free. A gleam of sunshine made the raindrops on the drenched leaves sparkle and twinkle. I left the clump and suddenly came upon a hut, guarded by a dog, which barked loudly. His master thereupon appeared from the other end of the hut. I inquired for the shepherd, but learning that he was far away on the next hill I told the rabbit-catcher[2] (as he proved to be) about the sheep in the hole. 'Damn my eyes an' ears!' he exclaimed, 'I arn't never sin such a thing! Let me back my dinner-pot a bit, an' I'll come along.'

He was soon ready, and accompanied me to the hole. I do not know what he expected to see, but when I led him to the actual spot he was amazed, and exclaimed, 'Damn my eyes an' body if I ever see sich a thing before! It'll be a tidy job a-gettin' 'em out; we shall want another man from t'farm.' Being minus a foot he could not go as fast as I, so while he went back to his hut for tools I walked on to the farm.

The carter started off with a rope while I went to the farm-house, but the family were at chapel, and by the time I got back to the hole again the four sheep had just been dug out and freed. Then we all had our second surprise, for in a little deep 'pocket' was squeezed a fifth sheep, which had apparently tried to get out of the way of the rest. It was quite a long job to dig round the poor creature, to rope it, and haul it out of the hole. The sky was dull again, but I used the only film in the camera to record the release of this last prisoner, which soon joined the others on the hill-side. I went back to the hole and in it found a small flint chisel, but the picture of the roped sheep was the best souvenir of the ramble.

The rabbit-catcher pressed me for my name and address, which I gave him. He tucked the card in his pocket and said he should give it to the farmer. 'Reckon 'e ought to mek you a handsome present,' he said, 'fur you saved him 'bout fifteen pound 'smarnin'. I never heard from the farmer, however. Being a 'chapel man' perhaps he thought I ought not to have been poking about the countryside on a Sunday morning. If he did not care about the sheep I should have been very pleased to have had them, but I had a small reward, for I discovered an ancient flint pick on the site of an old excavation on the next brow.[3] My real reward, however, came at a later date, when, on my way to look at the hole once more, I found a very fine flint axe seven and a half inches in length.

Airplanes on the Downs

I chose a bright February morning to visit a shepherd at Pyecombe, and was directed over the golf links near Clayton Mills.

The birds made me later than I should have been, for a party of long-tailed tits, working a hedgerow, kept me dodging and watching, and a flock of redwings delayed me again.

Retracing my steps, I found a few red dead-nettles and a few early celandines, and who could pass, quickly, the first celandines of the year? I could not, for the clean, fresh look of the plants, the brilliant golden-yellow cups, which are so shiny inside that they appear as though varnished, forbid any haste on the part of the finder.

I discovered the shepherd in a barn, and in course of a

chat he told me of recent trouble caused by some low-flying planes. I give particulars of the occurrence as related to me.

A minute before the arrival of the planes the countryside had worn its usual peaceful look. Here the links seem to interfere less with the general aspect of the landscape than at some other places, for the players start within sight of two old windmills; they pass the barn and sheep-fold; they hear the bleating of the flock and the notes of the old bells.

The yard of the barn was littered for lambing, and a few ewes with their early babies were already in pens in the shelter. The flock of heavy ewes were nibbling quietly on the hill, when suddenly the whirr of approaching planes (which is scarcely noticed in the ordinary way) became very loud. The ewes stopped feeding and looked towards the machines. Down came the planes, twisting, turning, and sailing low over the ground and over the flock. Instantly all was confusion. The terrified ewes rushed hither and thither, heedless of everything. The shepherd's voice was drowned by the noise from the engines, and fear of the dog was forgotten in the excitement, until the great dread monsters rose again and flew away, leaving behind a huddled mass of panting animals.

'I knowed how 'twould be,' said the shepherd. 'I diddun' have to wait long; they planes comed on Tuesday an' to-day's Sunday, an' there be four dead lambs come already; an' I be in fur more of it yet, I reckon!'

After I had taken advantage of the bright sun to snap him with two of his new babies – one white and one black – he again referred to the worrying subject, but I was really surprised to note how reasonable he was in his arguments regarding the affair.

'There be on'y one thing to do,' he said, 'an' thet is to write to the paper 'bout it. Stan's to reason them cheps in t'flyin' machines diddun' do it on purpose; I doan' say thet

– I arn't so silly! – 'cos there be lots o' things as happens to ye fur the first time afore you knows they could happen at all! Arter thet you looks out it doan' happen to ye again. If these men was *telled* 'bout it they wouldn' do it; but *I* can't write to t'paper! You see, there mus' be *somebody* as knows what flyin' machines comed here on Tuesday. There's soldiers an' sailors an' p'licemen an' sich, an' there's things they doos an' things they mun't do. It jus' wants some one to write it all out for 'em, an' p'raps before nex' October it'll be printed on their papers, so they knows not to do it.'

On a subsequent visit the shepherd reported that more dead lambs arrived, in addition to other troubles, and asked if I had 'writ to the paper about it' for him. I confessed that although I had never done such a thing before, I did write to a big London daily, but my letter passed unnoticed.

Perhaps, in the future, all pilots of planes will be 'telled about it', as the shepherd said; but if any of them could have seen the little dead bodies in the sheep-fold, and listened to the shepherd's sad tale, they would never require an official instruction on the matter.

The Flood

On the thirteenth of March I set out in my usual style to seek for beauty and romance across the countryside. The wonderful early morning light gave a peculiar brightness to the colour effects, as I passed furrowed fields and sown fields separated by faintly tinted hedges and lines of elms. The latter were already beautiful at a distance, with that elusive blush of smoky pink which appears to wrap their branches before the bloom actually gives the trees a distinct colour. Smoke from the chimney of a thatched cottage appeared as a blue vapour as it wreathed and ascended and faded away among some elm branches. Rooks were much in evidence; so were their companions the gulls. One becomes accustomed to seeing gulls inland as well as on the shore.

I did not find the shepherd I sought where his lambing fold was pitched last time, and on enquiry learned that he was yet another mile away. It was no hardship to do an extra mile – such a mile! – with countless things to see. First another shepherd's family of horned sheep whose white babies were capering and prancing as only lambs can do; next a sallow bush with buds of silver; then a party of chaffinches, not yet disbanded for the nesting season; then a patch of ivy of wonderful beauty, tinted with crimson colour. Interest without end! – hazel catkins above, dog's mercury below, and, lower still, hart's-tongue ferns in the ditch; a robin, a rabbit, a startled ringdove flapping, a clump of speedwell blossoms, a weasel, a bunch of cattle grazing, and a baby holly tree showing among some withered grasses. A tomtit was perched on the post of a farm gate, calling with shrill, sawing note, and warning me that there was no admittance that way, as a printed notice stated. He did not know that I was a privileged person with a passport to the ancient barn in the distance.

As I made my way toward the dike-bounded fields I could see hurdles in a mangold patch, and found the fold tenanted by a number of ewes and babies. Even to my eyes they appeared a somewhat poor lot of Southdowns, and it was not until I met the shepherd that I knew the reason for their draggled appearance.

'You be thinkin' things 'bout they fine lot Southdowns, reckon!' said the man with the crook dryly as we met. 'Come down t'barn an' see the rest of 'em!' On the way he told me how one of the dikes into which the yard drained had become dammed by a fallen bank. This caused the water to wash back into the yard, and as the reason for the trouble could not be traced at once, the place was soon flooded. It was saddening to listen to the account of the poor ewes waiting to lamb with water washing round them, and of newly born lambs in the same plight till they could be rescued. It took a long time to move over three hundred sopped ewes from the straw litter. A scout was sent out, who reported that the dike was clear and the sluice clear; consequently it was not until the shepherd himself left everything to search about that the fallen bank was discovered. The scout's ears must have burned when the shepherd spoke of him to me; but there was excuse for hasty words, for although by the time I arrived the water had receded and the place had been re-littered, the sad results of the catastrophe were still apparent.

In the ordinary course of events the stately old barn would have formed a perfect setting for a model fold. The main building, with its big old doors, through which the corn was carried in bygone days, faces the north, and the east and west 'wings,' with a connecting wall, enclose a yard only separated from the sea by one long field. This sun-trap was arranged for the main lambing fold. Inside the wings were rows of pens, now mostly occupied by ewes with their sons and daughters; but there were many mangers fitted with chains, and the shepherd informed me that the barn was mostly used to house fat bullocks. Just now the mangers were filled with fresh straw and fodder, which the good farmer had rushed down with all possible speed, while piles of roots and hay were stored in the centre of the big barn.

The shepherd gave me a surprise. He opened a connecting door, and there, in a temporary stall, was a cow, to provide the necessary milk for so many bottle babies and weak ewes. It was the first time I had seen a cow installed for this purpose.

It was only when we made a tour of inspection round the pens that I realized how deeply the shepherd was affected by the upset. Some of his sheep had been with him two years, some three; now a number of them were dead and many ailing, while many of the first lambs were born to a life of hours only, after all the months of care which had been expended on the ewes. The shepherd remarked on each pen in turn. 'This ol' lady lost her lamb!' 'This 'un had two – on'y one left now!' 'Two more for t'bottle – no milk, you see!' 'Here be 'nother! If I get her over to-night she'll live.' 'This un's bad! I've done all I can, but she'll got out, reckon!' And so on. A weary list indeed, only softened by the endearing phrases whispered to each patient in turn, which I was not supposed to notice. The poor man seemed quite crushed by his misfortune, for this little flood, not worth a newspaper paragraph as news, was as bad to him as a shipwreck.

We worked round to the centre again and my friend cut up many roots in generous slices. As these fell into a large trug placed on a wheelbarrow a soft 'Moo!' sounded through the partition. 'All right, Pansy. I hear ye!' he called, and in due course the gentle cow received her midday meal (and a caress as well), after which he led the way to his hut, where the stove needed stoking in readiness for the warming of the milk on which so many frail lives depended. Here I left my friend to his duties, for somehow I felt that I could not intrude upon the patient efforts of a beaten man who would not yet own defeat.

I set out this morning to seek for beauty, and found it in the elms, the smoke, the silver sallows, the crimson ivy, and all the rest. I also found sorrow – hard and crushing sorrow; yet in the shepherd's tender devotion to his poor unfortunate family surely there was beauty too!

— DOWNLAND NATURE — NOTES

Cissbury Ring in June

There are times when certain things become very tiresome; when the difficult problem of making two ends meet is made still harder by unwelcome noises and interruptions. Then the thump of your door knocker, the blare of a street band, the neighbour's gramophone and yelping dog all combined may prove blessings in disguise if they only compel you to leave everything for a few hours and go out on the Downs.

There, whatever the time of year it be, a wonderful peace awaits you, if you will let the fairies lead you to their haunts.

In early spring the treasures we find are mostly in small quantities, but now, on this fine June day, bewildering varieties of beauty are before us; masses and heaps and piles of beauty and sheets of colour, each one of which is composed of a thousand smaller bits that call for minute inspection.

As I stand here on Cissbury Ring,[1] near the rampart, baby rabbits frisk about the limestone slope and play among the clumps of golden stonecrop. One of them just appeared on a mound and came face to face with a wheatear, to its great surprise.

Other birds are in evidence. Several linnets are flying round. A yellow-hammer is calling from the top of an elder, and I can hear a skylark somewhere. In the plantation to the left ringdoves are hidden in the beeches. I shall find them presently. Meanwhile a wren is busy round the root of an elder bush growing in one of the pits.

Something fresh meets my eyes every minute as I wander along – a clump of burdock, a stunted yew tree, a ferret collar and line[2] hanging on a gnarled hawthorn, a cinnibar moth in gorgeous pink and brown dress, flickering about over the turf, a long trail of traveller's joy, and a blue butterfly. I cannot hurry a step. I think the same as a tramp, who once said to me, 'Ah reckon all der wunnerful things as be about zummer time be 'nuf ter maze ye!'

A glance over the rampart at one point gives a view of the glittering sea in the distance, and on the farm below a flock of sheep moving – the same flock that I visited at lambing-time. Further on the wonderful view of Downland compels one to stop and admire it without haste.

Furze silhouette. Barclay Wills' pencil and black wash shadow drawing of a sprig of gorse

A white flake of flint lies on the earth, shot out of a rabbit burrow. It has been trimmed a little and one edge is exceedingly sharp. It was probably a knife in the Stone Age. I pocket it quickly, for my attention is taken by two Holly Blue butterflies. These are common in the neighbourhood – they are even to be found in the streets in Worthing – but they never fail to attract me. Small Heaths and Meadow Browns also flutter about as I make my way round. At last I find the ringdoves in the beeches and oaks, also a pair of turtle doves and a green woodpecker. I leave the plantation and once more rest among scented thyme and trefoil and devils-bit scabious and beautiful grasses, and a bumble-bee is pleasant company. At last I leave it in possession and wander quietly along a hedgerow. Here, through stopping to watch the tumbling antics of a pair of peewits, whose plaintive cries attract me, I also discover a hare feeding. My glass brings puss very near. Presently I retrace my steps a little way. In parts of the golfer's domain the hay is being carted, but where I am the grass is far above my knees. The path goes downhill all the way. On the left, at the edge of the cut grass, five magpies are feeding, while tireless larks are singing overhead and their joyous songs follow me until I reach the highway.

What a ramble! What beauty crammed into three hours of liberty! Yet every one can do the same and take their share. Go to Cissbury on a fine June day, such as this, and if you can honestly say afterwards that you are no better for it, there is no hope for you! You are only fit to stay indoors and slop about in old slippers. You are only fit to enjoy your full share of the knocker's thump, the dog's howl, the gurgle of the gramophone, and the blare of the street band and other awful noises for ever and ever and ever.

July in Downland

To-day I lunched in the open, in a sheltered Downland bottom. From the high brow above me, where a dewpond reflects blue sky and flying birds, I might have seen the distant telegraph poles along a road leading to Ditchling Beacon. Here, however, was a fairy kingdom a hundred miles from telegraph poles, and not a thing in sight to suggest that there had been any change in general effect for a century or more. Only in one other place (in the depths of the New Forest[3]) had I experienced, to the same extent, that wonderful peace which pervades an ancient solitary spot – where time is not measured and Nature reigns supreme.

My seat was a mound of heather and thyme. Another mound made a cushion for my back. My carpet was of turf and Downland flowers, of which I counted fourteen varieties within reach of my hands. Before me grazed a flock of sheep and lambs, moving slowly towards me, old and young calling to each other in a bewildering variety of tones. Mixed with their voices was wafted the sound of their ancient bells, and these formed my orchestra – the same soft and pleasing one which had played for a hundred years or more.

A pair of stonechats flitted from bush to bush in a furze clump near by, and while I sat I was visited by three linnets, a yellow-hammer, a wheatear, a bumble-bee, and a Clifton Blue butterfly.[4] Just as my meal was over the butterfly returned. I watched it as it visited many blossoms, and at last the winged gem alighted on my boot and spread its brilliant wings in the sunshine. One by one the sheep gradually moved past me up the slope towards the brow. They disturbed the blue butterfly. There were some 'cluck' bells among the canisters, and, as the sheep ascended, the intermittent notes from each bell grew softer. I tried to find words for those notes, as some try to do for bird songs, but it could not be done. They called down to me from the brow. I did not wish to move. I became, for the time being, a part of the mound, satisfied just to be there, to bask in the sun, to inhale the scent of the bruised thyme and listen to the twittering of the linnets in the furze. For an hour I reclined, day-dreaming, till I dozed, held captive by the spell of Downland.

Presently my friend the shepherd[5] appeared, with his dog, coming slowly over the turf – no interruption, this, to break the spell, for he belonged to the scene and was part of it. He joined me on the mound, and after a chat we 'went for a doddle,' gradually turning the sheep towards the bottom again. Then we sauntered along, as his team were driven slowly towards the fold. At a gate leading to some higher ground where the fold was pitched they were methodically counted, and numbered five hundred and twenty-six. The fold was on a fresh piece of ground, full of mixed feed, which the shepherd termed 'rubbidge.' We stood for some time outside the hurdles, watching the flock. The dog lay at our feet. Suddenly a fine magpie flew into the fold, and perched on the back of a sheep. Then, as suddenly, it dropped to the ground, and, a moment later, reappeared on the sheep with a beetle in its beak, with which it flew away. It was soon back again, however, and we watched it at its occupation for some time. The shepherd told me that it had lately been each day and often brought beetles on to the backs of the sheep and ate them there. This struck me as a most interesting and unusual occurrence, and I was glad to have had the pleasure of watching it. Usually the magpie took little notice of the shepherd, but no doubt knew me to be a stranger or I should have seen more. I found that the beetles in the herbage about the fold were dor beetles.

Colour in the Cornfield

Hidden behind the thick hawthorn hedge, which is guarded by an army of stinging nettles, is a little cornfield, though only the inquisitive rambler, who treads down the noxious nettles to look over a little gap, would know of the beauty concealed there. Summer grasses line the inside of the hedge; already they are as tall as the cornstalks. Next to the grasses a narrow space allows one to walk slowly along the edge of the field or to bend down and peer at the fairyland scene. Among the green maze are splashes of scarlet and spots of blue – a riot of colour – for here grow poppies and cornflowers[6] side by side.

Unless you crouch down and keep still the mass of stems seems but a patch of green tints, but kneel there for a minute and the principal plants are revealed in detail. The stalks of the wheat are thin and stiff, but the massed effect of light-green and dark-green patchwork is caused by their ribbon-like leaves, which throw bands of shadow across each other. Tall poppy-stalks hold jagged leaves and long, hairy flower-stems topped with scarlet cups or fat, green buds. Here and there are bursting buds with their little two-piece caps still clinging to the scarlet petals as if unwilling to drop.

The tall, sturdy stalks of the cornflowers have shoots and leaves at intervals all the way up. Very narrow leaves they are, with one side green and the other silvery grey. The stems end in small buds.

The youngest are hard knobs, others are almost like knapweed buds topped with a blue spike; some, older still, show a few ragged petals stretching out, but the glory of the field is in the freshly opened blooms, shallow saucers made of rings of vivid blue stars. So soon do they come to maturity, so soon do the wonderful, compact blue heads change to loose, pale ghosts of their former selves that it is worth more than one journey to the flowers to capture the beauty of the first blue heads in their prime. Only the youngest and freshest are worth picking to take home.

The big, fat buds of the poppies are very fascinating. A two-piece cap parts suddenly and drops before us. It is stiff and papery; at our touch it crackles or gives a rustling sound as does a one-day growth on a man's chin. The big bud remains, like a blob of red sealing-wax, but presently the petals unfold and droop, and for a time the young flower hangs on its stem like a little skirt of crinkled silk. Gradually the little petals develop and expand, the flower opens and lifts its face toward the sky. A bunch of these beautiful buds is easily carried home inside a bundle of grass; it is quite worth while to bring them!

The Wren, and Some Others

From boyhood I have loved wrens, I know not why. I cannot explain the fact any more than I can say why I detest house-spiders, castor-oil, and the little piece of fat in the centre of a leg of mutton; but, as they say in Sussex, there 'tis.

Much of the wren's attractiveness comes from its trim appearance and its incessant and peculiar motions. Watch it as it drops from its perch – it is as if the plump little creature were lifted ever so gently and wafted down. Watch it again as it threads its way through any tangle; it does not appear to hop nor to creep; it just progresses and passes through each little archway without any fuss or apparent effort. Sometimes it ascends a bole after the manner of a tree-creeper, yet not like it, nor yet like a mouse. So gently does it move that a woodpecker, beautiful as it is, appears an ungainly clockwork creature by comparison.

A bramble against the side of a chalk pit or a fruit tree trained against a cottage wall offers a very pretty stage setting for a nut-brown wren. Then its antics are a real joy to the bird-lover. Its slim bill points upward, downward, sideways, as the little bright eyes spy into every crevice.

A page of adjectives would be needed to give the wren its due from me, and I am always glad when I find someone else who appreciates my favourite.

As I stood with a shepherd inside his lambing-fold a wren appeared on one of the hurdles and instantly dropped down to the foot of the nearest hay cage and dodged about among the straw litter. 'Thar she be,' said the shepherd, 'back agin! Sims as if she can't kip away.' I told him how fond I am of the little brown bird and asked him if he were fond of it too. 'Yes, I be,' he replied, 'fur she be summat like my wife, if you unnerstan' my meanin' – busy, an' flittin' 'bout quiet like, an' nara bit o' fuss. Why, I'd on'y jes' putt out t' lantern 's marnin' when she comed out t' rick where she sleeps nights. I saved 'er a bit o' bacon, but I putt it down a minnit on t' top step' (the ladder steps of his hut) 'an' Mas'r Bobbie found it! – 'e be a sharp 'un an' no mistake!'

Our conversation drifted from one bird to another and their visits to the fold. He liked the shufflewings[7] ''cos they be quiet.' Chaffinches and sparrows he reckoned 'a bit of a noosance,' but the stonechat or 'furze-Jack' was more welcome because he was 'summat like Mas'r Bobbie,' the robin. 'Wagtails,' he said, 'is nice birds, but they be a bit fussy, runnin' an' waggin', all wound up like, but they do ketch t' flies, an' no mistake, so I saz to 'em, 'Ketch t' flies, ketch t' flies,' an' they doos; damn all flies, I saz, fur I can't abide 'em. Reckon t' goldfinch be a purty bird,' he continued, 'but they doan't often come in t' fold; they likes the thistles.' (He could not pronounce the word properly, and said 'sizzles.') Then I mentioned starlings. 'Oh, starlin's,' he remarked, 'I doan't min' starlin's, fur they doos their dooty, an' t' ship like 'em.'

There seemed to be one visitor to the fold whose beauty failed to impress the shepherd. This was the yellow-hammer. 'Times I likes 'em, an' times they fidgets me,' he owned. He admired the cock birds for their beautiful colour, especially those with pure yellow heads, but as they usually visited the fold in the company of chaffinches and sparrows to pick up what they could they were bracketed with these as 'a bit of a noosance.' He told me how a yellow-hammer had unwittingly upset him. It was on a baking summer afternoon, when the heat was so oppressive that the sheep panted, yet a yellow-hammer, perched on top of a bush, disregarded it. 'I wur very near t' bush,' said the shepherd, 'an' I warn't feelin' well as I do now, an' thet thar bird kep' sayin', 'Bit o' bread an' no cheese, bit o' bread an' no cheese,' till I wur fair sick o' listenin' to 'un, so I throwed a stone through t' bush an' froughten 'un away. But I wur paid out, fur it wur a shepherd's crown I throwed what I'd picked up, an' arterwards I couldn' fin' it nohow, an' it wur a good 'un. I ain't foun' nothen' since, an' t'ent likely I shall till I finds 'nother crown! When I does I wun't lose 'un, not fur any ole bird.'

Dartford Warblers

It is quite an unusual occurrence now for any of us to return from a ramble and say that we have seen a Dartford warbler. It is still more unusual to be able to record an intimate acquaintance with the little bird. The month of May is the best time to look for it, as its restless antics during the breeding season sometimes lead to its discovery in patches of furze which have appeared untenanted to other times.

It is no wonder that W.H. Hudson christened it a 'fairy,' and loved it as he did, for its attractive appearance and its incessant and varied motions compel us to stand and watch it, and to linger in order that we may do so again.

It was delightful to find that the thick furze within sight of a certain mill sheltered a pair of Dartford warblers. On this May morning I spied them beneath the shadow of a great gold-splashed clump. Their long tails were elevated. The male in his rich, dark dress, donned for courting, hopped to and fro incessantly before his sweetheart, with his head feathers raised into a tiny crest. He talked to her many times with sharp, chattering note, which I heard very plainly, for I was near enough to watch his bill open and shut as he called. Away flew the sombrely-dressed lady, away he flew after her, and away I went in pursuit.

After looking in vain for a long time I saw him on the tip-top spray of a bush preening himself. Now he perched rather upright with tail depressed and arranged his breast feathers with his bill, then suddenly, after a vigorous shake, he spun round and darted into the air. He hovered for a moment with tail spread, calling as he did so, then he returned to his perch, but instantly quitted it for a lower one. Gradually he worked downwards and entered the clump. I crept near and could hear him inside. He called continuously, and his monotonous and somewhat harsh little note, given once, and sometimes twice, sounded impatient, and almost like an exclamation of discontent at not finding what he expected. This note would not have attracted attention at any distance, and it sounded very weak against the twittering of linnets which were there.

After a few minutes his little sweetheart came to find him and lured him away once more. I wondered what food he could have found, and by beating the bush over a handkerchief obtained a surprising collection of tiny creatures, which proved a sufficient answer to my question.

Perhaps I shall find him again another day! I would walk miles (as any enthusiast would) to watch the 'feathered fairy,' for when he is in evidence on the furze spikes his antics make up for all the quiet time that he spends skulking below.

The Dartford Warbler. Barclay Wills' superb illustration of this Downland bird was based on close observation of the bird in its wild state. A surviving sketch ('Notes made from a dead bird shown to me in a taxidermist's shop') shows, however, that he also examined specimens

The Stonechat

So many of our rambles in Downland lead us near or through patches of furze that we must, sometimes, take particular notice of the stonechat. The bird is so pretty, with its black head and rusty-red breast, with its dark body and tail, and white collar and wing patches, that it is attractive, even from a distance, although our binoculars reveal its full beauty.

During my visits to Sussex sheep-folds in the spring I have generally seen some stonechats. The hurdles and furze and the coops all offered excellent perches, from which the birds darted into the air after passing insects, or down among the sheep, to capture flies and other creatures. Although the bird may be looked for about farms, stack-yards, commons and waste places, it is more often to be found among the furze. Perched on a tip-top spike or twig, it flirts its tail and calls to us as we pass – whit, tack-tack – whit, tack-tack – but as soon as we approach it drops down and flies along, near the ground, and re-appears at a short distance, ready to repeat the performance.

The stonechat lays four, five or six eggs – greenish-blue, spotted with brown at the larger end – and makes a rather large nest. This is built on or near the ground among rough herbage or brambles, or at the foot of a thick furze bush. In any case it is usually securely hidden, and it often takes us quite a long time to find it, even when the actions of the owners prove that we are near to it. In my patient endeavour to discover a nest in a furze patch in Dorset, I located a new haunt of the Dartford warbler, and I was very glad that I had responded to the stonechat's invitation to stop and admire him.

The Stonechat: 'Perched on a tip-top twig, it flirts its tail and calls to us as we pass.' One of two preliminary drawings of the bird among the sketches of Barclay Wills. The other, earlier, sketch is dated 'Bournemouth Chine December 26th 1903'. The word 'sold' on this, 1904, version refers to the finished picture, not the sketch

The Wheatear

The wheatear is a summer visitor to this country. It arrives in March and stays until August or September, and, although it is widely distributed, the South Downs offer so many 'desirable building sites' on its arrival that the bird is a familiar one in Sussex. The nest is usually built in crannies such as are found in old chalk-pits or stony places, and occasionally in rabbit burrows; consequently, the bird is mostly in evidence on or near the ground.

A male wheatear, in his summer plumage of buff, grey, black, white and brown, is a beautiful bird. When seen against a dark background, the plumage appears very light, while the head seems cut in two by the black eye-stripes, and the body by the dark wings. Its white rump feathers are very noticeable as it flits along in front of us, but when it alights on the ground its wings are generally crossed over its tail and these feathers are hidden. The white plumage under the tail is then the brightest part of the bird.

On ploughed ground it has a habit of flying short distances and alighting suddenly on the tops of the clods. Then it draws itself up very erect for a moment, and as its dark legs do not show at a distance, it appears to be fixed on top of an invisible wire.

Whether sitting on a post or rail, wagging its tail, or rising in the air after insects, or quartering the ploughed ground, or skimming along over the Downland turf, the charm of our pretty and interesting visitor always attracts us, and we rejoice that the horrible custom of trapping wheatears for food has, happily, nearly died out.

The Wheatear. Barclay Wills' black and white painting of this once common Downland bird

The Salvington Goldfinch

On July 11th I was on the top of the ridge by Salvington Mill, looking down the bank towards the Findon Road.

In the furze bushes behind, the pods were snapping as they ripened in the heat. Burnet moths were flying round, conspicuous by reason of their bright colour and peculiar flight, while skylarks and meadow pipits were feeding on the bank.

Seated there I used the binoculars idly, observing dear old Cissbury Ring, the farm below, and the valley between the hills. On a bush down the hillside the glasses located two goldfinches, and while watching them one rose in the air and flew rapidly in circles above its mate on the bush, twittering all the while. As I looked down on the bird from my seat above him the white and yellow markings on his spread wings and tail gave a most extraordinary and beautiful effect. Movement and markings combined produced an impression which seemed slightly familiar, but it was not until the bird had repeated the performance and returned to the bush for the third time, and then followed his partner to a distance, that I captured the fleeting idea and realized that a very similar effect is produced by a large humming top, when it starts in large circles.

Was this a usual event in the life of a goldfinch? I have not observed it before, nor since.

The Green Hairstreak Butterfly

It is a common occurrence for vivid recollections of pleasant hours in the past to be recalled by the sight of some flower or other object, and such was my experience to-day.

I lay idly among the thyme on a Falmer down examining the plants near me with a large magnifying glass – a little habit which is conducive to wonder and reverent appreciation – when a little butterfly (a Dingy Skipper) flitted by. Presently more followed, and I tried to get the glass over one as so many seemed inclined to settle at one particular spot near me. I had barely done so when another flutterer appeared, and instantly and last twenty years of my life melted away, for on the thyme stood a little winged gem, a Green Hairstreak.

I knew that its hidden upper surface[8] was only snuffy brown, but the gorgeous beauty of its green under surface, seen through the magnifier, stirred my enthusiasm again and recalled the day I first made the butterfly's acquaintance in the extreme north of Sussex.

It was in a woodland lane bordered by beech trees that I saw a Green Hairstreak flit by. I followed instantly, net in hand, tingling with excitement, for here was a fresh species to sketch if I could capture it alive! The nimble creature turned up a glade and I went after it, dodging from side to side to the end of the glade, for I felt that I could not afford to lose it through a false stroke. Suddenly, with that rapidity of thought which is so usual in times of crises, I realized that

I was an intruder in the glade. A pretty girl was resting against a beech bole. A young man was lying with his head in her lap and her fingers were in his dark, curly hair. Instinctively, I moved to and fro on tiptoe. Etiquette demanded that I should retire, but the butterfly lured me on nearly to their feet. As I approached, the girl's hand covered the man's ear. I saw her do it although I was watching the butterfly. Then I made one of the neatest strokes I had ever made and captured my prize. I looked at the girl, who surveyed me with twinkling brown eyes – such expressive eyes! – they changed and almost implored me to go quietly away while her hand still covered her companion's ear. Very gingerly I stepped backwards a few yards. Her roguish smile was my reward. As I turned, her hand left the man's ear and her other hand covered his eyes as she waved farewell. It is no wonder that I have always remembered my first Green Hairstreak of Sussex.

Clouded Yellows

Just as we reached the fold I picked up the largest 'shepherd's crown' I have found, presumably a luck-bringer. It is certainly a souvenir of a lucky day for I had enjoyed the company of many Clouded-Yellow butterflies. Never before had I seen so many; never before had I had an opportunity to study them and watch them in flight time after time. This year (1928) is what is known as a 'Clouded-Yellow year',[9] when these beautiful golden-yellow creatures are abundant.

The first few that I saw were on the stubble among stooks of oats, and as they flew to and fro their quick movements deceived the eyes. It was not easy to mark down the spot where they had apparently settled. The rapid twinkling flutter of their golden wings reminded me of the movements of a bat – a restless, endless search, often over and over the same piece of ground – and as they turned and twisted their colour mingled with that of the yellow stubble. In another spot they were still more wonderful, for among the tangle and dried grasses on a bank, over which they flew, grew ragwort, scabious, knapweed, harebells, and a few late rampions, also bunches of dried, shining, silvery cups, which marked the forgotten blooms of the first knapweeds of the season.

On ragwort flowers the Clouded-Yellows were lost to view, but clinging for a moment, with closed wings, to the bending scabious heads they were delightful. Only once did one settle with opened wings – its perch was a harebell – it settled as if by mistake, and rose again almost at once; yet in those few seconds it made a fairy picture by the exquisite contract of colours. It was one of those rare, sweet moments which sometimes reward the rambler – a moment of time so filled with beauty that one is almost overcome by it; a moment of such emotion that the next is a bewildering realization of sudden return to ordinary things, as after a wonderful dream. The journey of hours, the tramping of miles is as nothing when you know that you have captured an impression which will never fade.

The Green Hairstreak, just one of the Downland jewels Barclay Wills observed

'On the thyme stood a little winged
gem, a Green Hairstreak. I knew that
its hidden upper surface was only
snuffy brown, but the gorgeous beauty
of its green under surface, seen through
the magnifier, stirred my enthusiasm
again.'

— SOMETHING OF SUSSEX —

Sussex Oxen

To Mr. H. Coppard, the shepherd at Water Hall Farm, Patcham, I am indebted for my first interest in Sussex oxen. He saved me some of the old ox shoes (or 'cues') which he found. As in many other places a team worked on this farm years ago, and their shoes are still turned up occasionally by the plough. Mr. Nelson Coppard also gave me more samples from Mary Farm, Falmer. After that traces of oxen often greeted me, especially at country forges. In another chapter I have recorded my visit to Mr. Hoather of Kingston, who gave me the last ox shoes he made and which were not used. These were larger than those seen in other places. If enough shoes were available it is probable that an interesting series could be made, showing the variation in the products of the various forges.

These old shoes are real links with the past, for the last ox team left in Sussex at this date are not shod.

Some of the oxen had their sharp horns fitted with metal knobs – a necessary precaution – to prevent injury to other members of the team. They were made of brass or some other hard mixture, and were screwed on the tip of the horn by means of a knobbing iron. The ridges on this tool were designed to clip into the slits cut in the knob, thus giving an excellent grip. The knob was turned round with the iron, the thread inside it cutting its way into the horn.

I was particularly pleased with this gift from Mr. Hoather, for a knobbing iron is now a curio, and it would be difficult to guess its use without some explanation.

My desire to see the last team of oxen[1] working in Sussex was gratified this morning (May 13th). Although in other places a wealth of buds and blossoms were waiting for me, yet the lure of Downland was in the air. Larks were singing, goldfinches were pairing, and as I made my way from Seaford to Exceat I met many old friends. Orange Tips flitted by, Holly Blues fluttered about the hedge which sheltered the wild arum lilies and white nettles.

At last I came towards the farm and crossed a small bridge. On the left was a hedge with a deep ditch behind, full of reeds, and here some sedge warblers were singing with great enthusiasm, as is their habit. Although I saw them again later I could not respond to their invitation just then, for on the opposite side of the narrow road, and separated from it by a dike, was a flat meadow, dotted with small cowslip blossoms, and there were four oxen pulling a large roller.

I was anxious to see their owner, Mr. Gorringe, and to

An ox being shod at Saddlescombe Farm, c. 1888. The ox's feet are held in position by being tied to a tripod, while the ox-boy sits on its neck to prevent it moving its head. The smith is seen fixing the cues, using nails that have been greased with pork fat. INSET The last team of oxen in Sussex at Exceat Farm. This four-ox team is seen in the charge of Mr. W.E. Wooler, who is holding a short goad. They are pulling a roller and are wearing nets to prevent them from grazing. An ox-yoke is visible on the neck of the leading pair

obtain permission to study them closely. This being done I was soon beside the team, chatting with their driver, Mr. W.E. Wooler.

We ate our frugal lunch together while the oxen stood by, for their meal was to come later. They were fitted with nose nets lest they should be tempted to graze while working on grass. It chanced that one of the six was unfit just then, so this made a compulsory holiday for his companion, but in the ordinary way three yoke were always seen. I had time to make a leisurely inspection of the big beasts, their yokes and chains, their nets and horn-knobs. From Mr. Wooler I learned the names of the oxen – Lamb and Leader, Pilot and Pedler, Quick and Nimble. They are steady and strong and are cheaper to work than horses. This team can plough a good acre and a half in a day. Their diet is hay, straw, and swedes in winter, and grass and cake in summer. They are fed first thing in the morning and at four o'clock in the afternoon, and work from seven till three.

The brief half-hour for lunch passed all too quickly. Mr. Wooler picked up his goad and waited for me to take a snapshot. He wore a cowslip blossom in his coat. The word was given and we stepped along by the leaders. Then I realized how that steady, regular pace, which appeared slower from a distance, would account for the work done by the team. Up and down the meadow we went. At each turn the goad was laid lightly over the necks of the leaders, and a murmured 'Ay' or 'Gee' was sufficient instruction.

The goad is a long hazel stick with a piece of wire in the tip of it. This tip had the same appearance as a lead pencil with a blunt point. The goad used by Mr. Wooler should really have been much longer, but it was not easy to obtain a long hazel rod at Exceat.

At last I left in order to visit the persistent sedge warblers, and after a long interview I turned to make my way over the hill by West Dean, where Mr. Dick Fowler carries his Kingston crook and tends his Southdown flock.

The oxen had just reached the edge of the meadow. 'Gee,' said the young Sussexian, and the big, patient beasts swung round and started another length.

NOTE

Since the above was written the Exceat oxen have been sold, and for the first time within living memory the farm is without a team. The sting of this sad record is lessened, however, by the fact that a new team of six has been started by Major Harding at Birling Farm, East Dean. They were broken in to the work by Mr. Wooler. When I heard the news I drank to the health of their owner, for it is no small thing to many of us that the chain of years is still unbroken and that we may continue to enjoy the sight of an ox team on the Sussex Downs.

The Hurdle Maker

The hurdle maker is not the least among the workers whom we meet in the shepherd's domain. He also is intimately connected with the flock, for he often acts as sheep-shearer in the season, while a large number of the hurdles made by him in the winter and spring are used by the shepherd at lambing-time for making cosy pens for ewes and for fencing. His well-made cages hold the sweet hay supplied for food at that time, so that he has a real interest in the result of his labour. He is among those very clever country workers whom we are so apt to overlook and take for granted. To many he is unknown as a craftsman, yet a visit to his little woodland clearing in the early spring, when copse-cutting is in progress, is a revelation.

I went into the wood to-day (March 19th), for I spied his little camp from the roadway. Among the litter of cut brushwood and chips the primrose buds among their clumps of tender young leaves appeared very bright.

The camp was protected from the wind by a screen of posts and crossed poles, banked up with faggots and hurdles. Two young oak trees were used as supports. In the lee of the screen the man worked at his craft. His open-air workshop was very compact. Between the screen and the little well-trodden path where he stood lay a supply of rods – some already split. A worn post of convenient height was fixed by the side of the path. Near the top a hole had been bored through it and a wooden peg fixed so that it protruded about three inches on each side. After a rod had been split at the thin end with his little adze on the head of the post it was opened in a fork round the post and forced forward. As the rod split the two halves glided over the pegs until they separated and fell to the ground.

Immediately opposite the post was the frame – a wooden beam about seven feet long, slightly curved and with a row of holes in it.[2] It was fixed securely to the earth by braces of split and twisted rods and pegs. In the holes stood the ten pointed hurdle uprights,[3] and the split rods were quickly selected and woven between. What strength of wrist the man had! The thickest end of the rod being fixed, the rest was dexterously wrenched round the outside upright, caught and woven between the others and forced into place by his leather-protected knee or his thick boot. The ends of the split rods were trimmed off neatly and quickly with a small hatchet.

The hurdle maker. He is holding his bill hook, standing by the sails of a newly constructed hurdle

His adze lay on the ground, also his tin of tobacco and box of matches. Near by, against a bundle of poles, on which his coat hung, was his umbrella – a big blue one, such as the shepherds use. This and a dinner bag completed his modest outfit.

Year after year he makes his hurdles and sheep cages. His hand is firm, his blows are true and made with that precision acquired by lifelong concentration. A young woodman working near by (who can also make hurdles, and owned that, after seven years' practice, he still has much to learn) paid a glowing tribute to the older man's skill and the quality of his work.

Ashurst Mill

It is good to tramp along the crests of the Downland hills; it is good to stand and gaze at the bewildering pattern of the Weald spread out before you. It is good to depart from your usual practice sometimes and descend into the valley to look back on the rugged sides of those hills over which you have passed. And while in the valley perhaps you wander a little way, or make a dash into this different country to see something that attracts you, for there are many treasures in the Weald, and among them are some windmills.

It was an autumn day – a morning of glorious sunshine after a spell of wind, rain, and hail, which had stripped trees and bushes of all the leaves that were ready to fall. I went to Steyning to find Ashurst Mill, and was anxious to arrive in time to photograph it while the sunlight lasted. The three miles of country road did not seem like three miles, for there was so much to see, such views and so many details to note. I saw rooks and starlings, a hovering hawk, a woodpecker, and a party of goldfinches. I met a gipsy family, then a man of commanding aspect in a great, wide-brimmed hat, driving a few cattle, then some roadside workers, all of whom spoke, as I passed, in that friendly country way which we miss so much in the towns. At many spots the full ditches, which apparently contribute to the local tributaries of the river Adur, were responsible for beautiful growths of many kinds, including graceful horsetails and green rushes.

At last I saw the old mill in the distance, but soon it was lost to view, and a big motor-lorry rushed by; then, as that vile creature disappeared and the road was clear once more, the mill appeared again, and I knew that I had found what I had hoped to find – a post mill in the open which could be sketched or photographed.

As I approached the roadside trees just before the mill, a tall white poplar, which reared its thin branches above the rest of the crowd, offered a pretty effect, for the silvery white under-surface of its dying leaves appeared like a mass of white blossom against the blue sky. On the bank below were many of the leaves. They had changed to a dull yellow speckled with brown, although the under-sides retained the white felt-like surface. These deeply lobed leaves and the smooth trunk suggested that the tree was younger than those sometimes seen with furrowed bark, although it was so tall.

From Mr. Etheridge, the owner of the mill, I learned that it has not worked for at least thirty years. He has not been up in it for twenty-five years, and it is unlikely that anybody will ever enter it again, for it is much decayed. One by one many bits of outside boarding and boards from the sweeps have fallen, one by one the ivy shoots have crept up the four

Ashurst Mill. An open trestle post mill, built in 1798, it had not been in use since the early 1900s and had fallen into disrepair when visited by Barclay Wills. He forecast that 'any gale may suddenly transform it into a mere ruin', and sadly the mill was felled by a gale in December 1929

brickwork piers, on which the timbers rest, until these have become green mounds. There is a tangle of growth under the raised floor, and an elder bush has pushed its way through the broad staircase. Never again will a miller tread the wooden steps! Already sixteen of them are embraced by a tangle of bramble stems, and the blackberries have ripened on this novel support. There are some bits of board among the bushes, and more look ready to drop.

Poor old mill! – soon it will even cease to be picturesque as it is now, for any gale may suddenly transform it into a mere ruin. Meanwhile, like the oldest inhabitant of a village, brave but frail, it lingers on, waiting for the inevitable crash, waiting for the day when all its friends will say to one another, 'So the old mill has gone at last!'

Nearly opposite the mill is an ugly, modern, red-brick cottage. It is useful as a signpost, for I was directed to turn down a lane by the side of it, and I was glad that I did so. What a lane! Only wide enough for one vehicle, it went wandering on and on. It forked, and I was careful to follow my instruction to miss the right-hand path (which leads back to Partridge Green) and keep to the left leading out near the Dial Post at Ashington. But I have not yet recorded the beauty of the lane. Although not one butterfly was using it, and very few birds, other than two marsh tits, met me, there was quite enough to see. The previous rough days had whipped away much of the foliage and strewn the weak weed-stems in all directions, so that the ditches, the little recesses, and hiding plants were all exposed. In some places the pools of flood water mirrored the branches above; in one a long line of sallows rose from the ditch – a line which will lure me back there in spring; but the glory of the lane was in the autumn tints, the broken masses, the dots and splashes of green, golden-yellow, orange, and brown, a variety that continued without ceasing all the way.

From Ashington one can walk to Washington, and there, on the left, you can climb to Chanctonbury and so back on to the downs.

You may be a lover of the hills and the trackways across them, but this lane is a pleasant change and a refuge from motor traffic. For those rambling in the neighbourhood of Ashington it is an alluring path, a beautiful approach to one of the most romantic and picturesque features of the Sussex landscape – an old post mill.

If I had Money!

It was quite a change from bells and flints to seek for an old wooden Sussex plough. Among the usual smaller items it seemed a large thing to hunt for. On each journey into Downland I made inquiries, but the result, from various causes, was always the same – no plough. Sometimes I heard of a place where one was used years ago; sometimes the advice given me was too vague to be useful, and it looked as if the expense of finding what I sought for would be more than the value of the plough.

One day I found Shepherd C——, whom I had not seen for some time, and on his advice visited a farm at Patcham, where he was known. To my surprise, his friend took me to the edge of some ploughed ground, and there, on the grass, lay the old thing I wanted.

Was it genuine? Was it complete? I asked, and learned that it was formerly drawn by oxen, and had been in use until lately. I needed no further proof, and my next query was as to whether the farmer would sell it, for my friends wanted it for exhibition. I met the farmer soon after, and he consented to my request.

Mr. Richardson, who had used the plough, was enthusiastic about it. He said he would rather have it than one of modern design. He regretted the accident through which the handles of his favourite had been snapped. He stated that after repair there would still be 'years of wear' in it, and that although four iron ploughs are in use these work best on level ground, whereas the heavy old wooden one is more suitable for the steep slopes on the farm.

His reminiscences of the oxen were very interesting. They were shod in the open in summer, and inside in bad weather, and two blacksmiths attended. It was not an easy job to throw them, for they understood what was to be done, and their dislike of the shoeing often caused them to tremble with nervousness when the blacksmiths appeared. When an ox was thrown its feet were fastened in a bunch, and some knack was needed to perform this little operation properly, so that any cramping of the limbs should be avoided. The shoeing was always accomplished as quickly as possible to release the animal from its unnatural position. Mr. Richardson generally sat on the animal's neck during the shoeing – a rather insecure perch, for he was sometimes lifted up by the movement of the head below him. He confirmed the fact about the nails being stuck into a cushion of fat pork, as described by the shepherd at Wilmington. He showed me some photos of oxen ploughing on the farm, also one of himself by a barn door with a flail in his hand.

I wonder how many men are left who have used flails and who are capable of teaching another how to do it? The thought brings others, and other queries.

Sometimes I wish that I had money, for there are so

many things that a rich man could do for Sussex while there is yet time to do them.

If I were wealthy I would build a wooden windmill on a Sussex hill to keep alive the memory of the old mills and millers for a few more generations, for there is opportunity, at present, to record every necessary detail and measurement from an existing example such as the one at Ditchling. My mill should be of oak throughout, and I would find the last of the old craftsmen to build it. My mill-house and my barn should be exact reproductions of the finest examples. On my farm a team of oxen should work, and a Southdown flock should charm the visitors with the music of old canister bells. The wagon horses too, carrying light loads and bringing corn to the mill, would wear their bells as in the old days.

In my forges old craftsmen should instruct others in the making of many things; the wooden plough for the oxen to pull should be fashioned there, and the last users of such things as flails should teach us all their little secrets. Quaint scenes, such as the shoeing of oxen would be seen again.

One more thing I would build – a real Sussex museum (not a home for foreign curios, for 'Aunt Jane's favourite parrot' and other such oddments so often thrust upon curators of local museums). Here the student should find all the records appertaining to the Sussex countryside, and round it many wild flowers should find a home where they could multiply in peace.

Alas! I cannot do it, although, as I tramp about, I note the many things which are fading out. There are plenty of men with money enough to do all this and more, and among them all there surely must be one with the heart to do it! Where is he? Is he afraid to take up a firm stand and do such work for the good of old Sussex? He need not be afraid, for he would have the support, not only of the members of Sussex societies, but of all the small people, such as myself, who are endeavouring in one humble way or another, to capture and record some of the vanishing beauty and romance of this wonderful county.

— THE PASSING OF THE — DOWNLAND SHEPHERDS

'This Hidden Sorrow'

It is strange and sad to reflect on the fact that the next generation will probably miss the joy of such old-fashioned company on a ramble as is recorded in the foregoing pages. Times and conditions are now so changed that the oldest shepherds and workers on the land are the only people who can inform us as to the details of many odd subjects; but their great experience, their broad outlook on life, and their love of thoroughness are already out of date. They are quite aware that they belong to a dying race, and while tending their flocks and wielding their pitching irons the ghosts of the past live with them.

I think that some of their peculiar indifference to other things than the job of the moment is due to the fact that the shepherd of to-day has lost much of his former importance. It was a sad day for the Sussex shepherds when the old type of Sussex farmer began to die out, for those who have come from other counties hold the farms but have not filled the places of the former owners. In many cases they and their employees are real 'furriners' to each other, and must always remain so; for although the new farmers[1] may have the best intentions their ideas are not in accordance with Sussex traditions. Their efforts are met with that most peculiar reserve which characterizes the natives of Sussex, and which is so difficult for those who come from other counties to understand.

Here are a few instances taken from notes of real interviews.

On observing to Shepherd P—— that some of his ewes were in very poor condition, he replied that the farmer had decreed that he should keep them for one more lambing. 'An' what is the use of that?' he said, with a gesture. 'I doan't want to look after ewes like that. The lambs wun't be any use when they comes. Boss bought wrong, you see. They sheep be no use for this farm. 'Twould pay him better to send 'em all to butcher an' start fresh. Them others are better now,' he said. 'I told him at the fair that they was the right sort, an' he bought fifty to try. When he sees the number of doubles I get nex' month he'll see I was right too!

Old Shep. William Shepherd, known as 'old Shep', was born at Duncton, near Petworth on 31 December 1849 and was probably the most photographed shepherd of his era. He also worked occasionally as a rabbit catcher, as shown here. When Barclay Wills met him, Old Shep bemoaned the problem of finding good beer: 'Time was when you could get a glass o' beer wi' a good head o' froth on it. That showed that John Barleycorn was in it! There is no Barleycorn in modern beer. It is almost pizen!'

My father was shepherd here,' he continued, 'an' I took his place; reckon I *ought* to know what sheep to buy!'

Shepherd B—— spoke of his new master from 'the sheeres'. The farmer rode about on horseback, and one day, after inspecting the flock, said to the shepherd, 'There is one of your bells lying near the track over the brow; I passed it yesterday.' I was not surprised when the old man said, with some warmth, that he reckoned the boss might have picked it up for him. Then he told me how he started off directly his work was done, and after a long walk and a long hunt came home with the bell. It proved to be one of his favourites. To him it was very precious, as any old Sussex farmer would have understood. The 'sheereman' passed it by; it was nothing to him; he did not ride round on his horse to pick up old bells!

I was doddling round with Shepherd F—— listening to his observations and little grumbles. He stopped suddenly and said, 'Here is the place where a pond was in th' old days. It wouldn' take much to dig 'un out agin, but this farmer wun't do it. He says if sheep want water take 'em down to t'other farm! But '*e* don't take 'em! *I* takes 'em an' *I* brings 'em back; an' time they be back from all thet way they be nigh thirsty agin, an' puffed. There be no boun's what he will say sometimes! One day he said as there was plenty rape. 'Yes,' I says, 'but 'tis not much good; the frostes have took all th' life out uv it.' 'What d'ye mean?' he says, 'by took all th' life out uv it?' I looked at him! 'You be the farmer,' I says, an' I walked off!'

It was Sunday tea-time in Shepherd N——'s cottage. The logs crackled while we sat round the table and enjoyed our meal. Shepherd N—— was off duty and dressed in his

Sunday clothes, and more 'at peace with all the world' than ever. Gently I led the conversation round to the subject of old Sussex farmers. The shepherd told us many things relating to old customs, and ended with a reference to the farmers of the present day in his own peculiar mixture of modern speech and old dialect. 'Dese noo blokes bean't de same as de old farmers,' he said. 'Mine doan' even unnerstan' why some sheep mus' be tended. I be sick an' tired o' tellin' 'e things, so now I jus says, "Yes" an' "No", an' all de time I thinks to myself, "Do as yer like, Mus' Farmer, an' get on with it!"'

'Bells?' he said, in answer to my query. 'No, *my* bells ain't goin' on *dem* sheep. I doan't like 'em, any more'n I like blokes like 'im!' Then he continued, 'All us ole shepherds be gettin' de same way now! – we've nigh give up grumblin'! Mos' of 'em is same as me, an' thinks to 'emselves, "All right. You get on with it, Mus' Farmer. You knows! You get on with it!"'

It is difficult to give a name to this hidden sorrow, this forlorn indifference, which has laid its hand on the shepherds of the old school, for a strange mixture of feelings it is. Once happy with a pittance and the good friendship of their employers they are now fading out, misunderstood by those who grudge even five shillings a day for the sole charge of a flock.

'Did you know old Jack Blank is dead?' said a shepherd to me, not long ago. ''E was one of the old 'uns! Thet mek's three this year! Us'll all be gone soon! He wur in thet pictur' I showed you of the old shearing gang. Well, I be the last now, all but one. The young 'uns wun't be like us old 'uns,' he continued, 'for everythin' be altered, you see – what wi' li'l flocks, an' bits o' wire, an' furriners on th' old farms. Ef a boy do start now as a shepherd he wun't putt up wid it for

long. 'Cos why? 'Tis the times, you see. Us old 'uns was poor, but things was different, an' a shepherd was *somebody* then; but now us be nobody at all! All the old bells be wearin' out too,' said the old man, 'an' nigh all on 'em 'a' bin mended; but *us* can't be mended an' 'ave noo tongues, like bells can, so us mus' goo!'

The shepherd stopped to put another log on his fire; then he said, 'Reckon 'twar a good idea to putt some on us an' t'pictures o' th' old bells int' a book, 'cos now 'twon't be all forgot; but it do amoose me, times, when I thinks 'bout you a-comin' a-ferretin' it all out!'

The 'Furriner'

With pleasant remembrances of a snug, well-appointed lambing-fold just visited I wandered along a track in East Sussex on a March afternoon. The sight of a small fold and a shepherd's hut with open door, just as a sudden snowfall started, was very welcome and I made a bee-line for the hut, where I took refuge with the shepherd for a few minutes.

As we talked the shepherd looked through the window. Suddenly he said, 'Sims I be wanted,' and went out into the open field. I naturally followed and waited while, with deft movements, he assisted a ewe. As her baby was born the snowflakes fell on it – a cold welcome, surely, for the little creature. The shepherd moved it, with the ewe, to the lee of a stack, for no cosy pen with a roof was available and no provision for comfort of any kind.

My queries on this point brought to light the old man's

grievance. His 'boss' is a hard man, a 'sheerman' (shireman) – a 'furriner' to the Sussex farm hands. He is from another far county (a foreign country, this, to the old man, who has never been out of Sussex). The good old days when a Sussex farmer owned the farm are gone for ever. The 'sheerman' will not have a lambing-fold outside his house although there is plenty of shed room. 'The open Down is good enough for that job,' he says. He is anxious to make the best possible price on his lambs, and welcomes the news of doubles (twins), yet he grudges the ewes their food. The old shepherd gets no allowance on the number of lambs reared, as he should do, but this fact weighs less with him than the harsh indifference of the farmer with regard to the flock. He told me that this spring hardly a bit of hay was allowed him, while his request for necessary straw and other things was not attended to. At the eleventh hour, in despair, he took the knife to a straw stack and carried away what he wanted. A double row of wattles with straw between in a line with the hay stack made a wind screen, but he had no helper and nothing more could be done for the first ewes, for their lambs arrived while the farmer dallied and grumbled.

During the one bright interval between storms of snow and rain I obtained a photograph of a white lamb (born just before I arrived) being washed by its mother. There was another in the fold born to the sound of a cluck bell, which the ewe was wearing, also an exceptionally small lamb which had been given shelter in the hut and fed from the bottle.

The shepherd did not wish to be included in the picture. 'Doan't take me,' he said, 'fur I be fair shamed to be seen 'ere – tho' it ain't my fault, as I told ye.' I could only agree, for the dismal appearance of the little place after each deluge was most depressing.

The old man's only remedy for the unsatisfactory state of affairs is to pack up and go, and herein lies the tragedy, for certain reasons cause him to cling to the cottage and the bit of Down which has been home to him for many years. That bit of Downland is next to nothing to the 'furriner,' but it is everything to the shepherd. Yet, one day, when his domestic affairs warrant a move he will visit an old Sussex farmer he knows, who will welcome him. He will pack his box and bits of furniture. He will collect up all his precious sheep bells and tie them up in a sack. Then, crook in hand, he will shut the garden gate, and he will say to the despised 'furriner' 'I be goin'!'

Where Golf Balls Stray

There is no apparent reason why I should play golf if I do not wish to do so. Equally there is no reason why I should pay a yearly subscription if I do not play, and yet I am supposed to tread a certain very uneven, muddy, puddly path, when, by vaulting the wire fence by my side, I can tramp on level turf. The puddle is the public footpath – I may stand in it as long as I wish; but although it is usual on country paths to find a second way round any boggy portion, such a thing must not happen when the wire fence encloses a golfers' sacred domain – in this case a vast space of open Downland, where the round-headed rampion (the 'pride of Sussex')[2] and other friends once flourished; where magpies still gamble with fate by strutting about on the putting greens.

To-day, as I approached the mud, however, I forgot the existence of the puddles, and splashed through them, for two butterflies (a Red Admiral and a Peacock) beckoned to me to hurry away from the links to the open Down. They lured me on for some distance to a big bed of downy thistles. There I lost them, but found a party of goldfinches. Still as a stone I sat, while the birds worked slowly towards me. The brown, dried thistles were made beautiful in their last days by the visit of the flock, whose dainty plumage sparkled in the sun. Their black and white parts flickered; the yellow feathers (brighter than the white) almost glittered, and the little red foreheads gave touches of warm colour to the scene. Their pleasant twittering was blended with the distant song of some old cluck bells, which I knew must be nearly a mile away.

On my way to find the sheep I stopped at the corner of a ploughed field to watch a hare squatting in a furrow. Something whizzed by my left ear with terrific force, and I jumped as a golf ball embedded itself in a clod. In my haste I cursed it and all its kind and stooped to pick it up, when, to my surprise, I found, just by the clod, a nice flint implement. By the time the golfer appeared I had forgiven him. I told him of the narrow escape of my ear and my discovery of the flint. He was a nice boy, and although he could not share my enthusiasm for 'flinting' as a pastime he apologized for his 'rotten shot' and we parted on good terms.

The ancient game of golf seemed to haunt me to-day. As I lunched with a shepherd on a hill-side I asked him for his opinion about it. 'I doan't reckon nothin' of it,' he said. 'It sims wrong t' fence in the Down puppus for sich rubbish.' Afterwards he offered a new line of thought on the subject. He explained in his quaint way that the fact of golfers being kept in their own territory made it safer and better on the rest of the Downs. 'I doan't want 'em all runnin' 'bout this yere part,' he said, 'I likes it quiet where I be. As to them players, there ent no accountin' fur 'em. The men be dressed up like gals an' the gals be dressed up like men. What wi' thet an' all they sticks to 'it a li'l ball, stan's t' reason yew can't mek' sinse of it. They be better shet up in their own fold away from I.'

This was certainly a new and refreshing point of view which I could appreciate, for a crowd of fifty people 'runnin' 'bout in this yere part' would have frightened away the whinchat from the furze spikes, and also the fox that trotted up a slope and turned round for a moment to inspect me before he slunk away into the nearest cover.

Over the hills I wandered until I came at last toward the old barn at Charman Dean. Seated at the top of the steep path which leads down to the bottom at the east of the barn, I watched a flock of sheep move along the valley below. Seen from above, at that height, individuals seemed to lose their shape and to become part of a great moving mass, which suddenly took the form of a gigantic flat creature creeping along. The leaders somehow formed a pointed head, odd members at the sides gave the idea of hidden limbs, and stragglers composed the long tail. It was only for a moment – there was no time to focus the camera, but the suggestion of a great antediluvian monster was forced upon one, for the illusion was perfect.

From the barn across to Cissbury Ring the ground is littered with flints. One never knows what may turn up in such a place, and if you start to look about the time passes unnoticed. The sun was setting by the time I reached Cissbury plantation and bringing the bright October day to a close.

While gathering a bunch of late honeysuckle bloom among the brambles I found an ancient flint axe lying on the

GLOSSARY OF SHEPHERDING AND DIALECT WORDS

Adze In the text a side adze. A small hand tool used by the hurdle maker for splitting hazel

Barrel Tubular end of crook attached to handle of ash or hazel

Black Ram Night A night of revelry with much singing, drinking and playing of games at a public house to celebrate the end of a shearing season

Bostal or Borstal A pathway up a hill, usually a steep one, such as those on the northern scarp of the Downs

Bottom A hollow or valley among the Downs

Bows *See* Sheep Bows

Cage Circular feeding frame made from split hazel, containing hay

Canister A sheep bell

Captain Head of shearing gang, distinguished by gold band worn around his hat

Catcher Member of shearing gang responsible for bringing sheep from pen or 'hopper' to shearers

Chin board Small piece of strong wood used for securing sheep bell around sheep's neck

Chummy Shepherds' felt hat, reputedly made from dog's hair

Claspknife Shepherds' general purpose knife

Clip A fleece

Cluck or clucket A sheep bell

Colt Junior member of shearing gang working in his first season

Combe Similar in meaning to 'bottom', but used more widely. 'The houses are in the hollows, the 'combes' or 'bottoms' as they are called . . .' (Richard Jefferies, *Wild Life in a Southern County*, 1879)

Coppice Underwood, normally hazel, but also ash and sweet chestnut. Divided into cants or sections and cut in annual rotation. Provided a variety of woodland craft products used by shepherds such as hurdles, feeding cages and thatching spars

Costrel   Small wooden barrel containing beer or cider

Covert   Small wood or thicket kept for game

Creep   Wooden frame between folds giving lambs, but not ewes, access to supplementary feed

Crib   Rectangular feeding frame containing hay, made from cleft ash

Cues   Ox-shoes. Also, spelt 'kews' or 'queues'

Dagging   The cleaning and trimming of a sheep's crutch

Dean   Another term for a Downland valley, particularly a wooded one. Now used only as part of a place name as in Rottingdean, Withdean, West Dean

Dewpond   Also known as sheep-ponds, mistponds, cloud-ponds and fog-ponds. A man-made, (usually) circular pond, lined with puddled clay, and filled by rain. The main source of drinking water on the Downs for sheep and cattle in most districts before 1939. Most have since been filled in or allowed to decay

Dipping   Statutorily required to take place after shearing to prevent sheep scab

Dirt knocker   Wooden mallet used for knocking out mud matted in fleece

Downs   (1) The South Downs, the range of chalk hills running westwards from Beachy Head, near Eastbourne, to the county boundary near South Harting; (2) More generally, any chalk downland; (3) Southdown sheep; (4) any other sheep breed traditional to downland areas, such as Dorset Downs

Drenching horn   Horn used for applying medicine to sick sheep

Droveway   Ancient tracks across the Downs used by Shepherds for moving sheep

Faggots   Bundles of brushwood sticks, also known as bavins and used as fuel primarily for firing ovens. Furze faggots, known as fuzzes were used for lining lambing folds

False tongues   Large pieces of thick leather fastened above the laces to keep out the wet

Folding   The practice of keeping sheep in enclosures of hurdles and one of the major characteristics of Downland sheep farming. The folds would be moved around the farm allowing the sheep to feed on root crops (mangolds and turnips) and other crops such as clover, mustard, rye and spring barley. In the summer the flock would be grazed on the Downs during the day and folded at night. Folding ensured the manuring of arable land and played an essential part in the planning of the farmer's crop rotation

Frail   A woven, flexible basket, worn over the shoulder by shepherds and other farm workers

Foot rot   A disease of sheep which makes them lame, if untreated

Furriner   A farmer, or other stranger, from outside of the County of Sussex. Usually, a term of contempt. *See* Sheerman

Furze   Gorse bushes. Also known as hoath or whin

Gangs   Group or Company of shearers made up of seasonal labour such as hurdle makers or old farm workers. They varied in size from just a few men up to about thirty. Gangs were named after the locality they came from such as Bury, Fulking or Clapham. A man could hand shear forty to fifty sheep per day

Gnomon   The sundial's pillar, whose shadow shows the time by its position on the marked surface

Goad   Hazel rod used for guiding oxen

Guide   Long thin end of the crook, bent at an angle to the barrel, terminating in a curl known as the whorl

Handbill   Hand tool with sharp blade used for cutting coppice, bushes or sharpening stakes

Hook   Another name for a crook, particularly those made at Pyecombe. A dipping hook had a longer handle to the crook with two opposing loops enabling the sheep's head to be ducked or held up as required

Horn lantern   Lantern with windows of beaten horn traditionally used by shepherds at lambing time

Hurdle   Woven panel of wattle hazel. Traditionally made 2ft 9in high x 6ft wide for Southdown Sheep, with a hole in the centre for carrying. A typical Downland farm between the Wars with 400 ewes would use 120 hurdles costing 1s. each. The gate hurdle made of cleft ash, chestnut or oak was also used extensively

Lambing fold   Wattle enclosure containing pens roofed with thatched hurdles and walls lined with straw or furzes. Built in the open often against threshed ricks providing protection against the weather. Adjoining fold(s) would be constructed to take the ewes and growing lambs. Alternatively, lambing would be undertaken in a sheltered yard or hovel adjoining a barn, sometimes with a small shepherd's room attached

Latten	A sheep bell
Landrail	Another name for the corncrake
Lieutenant	Captain's assistant in shearing gang, distinguished by a silver band worn around his hat
Lockyer	Peg made from bone or yew for holding bell straps to yoke or collar
Numbering or marking iron	Iron rod, dipped in hot pitch, used for branding
Oxen	Castrated cattle used on Sussex farms or in forestry up until the 1920s as draught animals. Breeds were either Welsh Black or Sussex cattle. The Sussex breed are a deep red colour and like the Southdown Sheep were improved by John Ellman during the late eighteenth century
Pitching iron	Iron bar with thick rounded point used for making stake holes during construction of fold
Pollard	The waste siftings from making flour
Puss	A female hare
Rape and rape greens	Not the now all-too-familiar oil seed rape which much Downland has since been ploughed up to cultivate, but either turnip tops or the plant *Brassica naps*, usually grown as food for sheep. Typically, Downland flocks were folded on to rape greens both to feed them and to ensure the manuring of the soil

Reeder	Rough leather washer used as bell tackle
Roller-wattle	Revolving pole on upright enabling sheep irritated by ticks to rub their backs and thus discourage them from rolling over

A roller wattle

Round frock	A smock of a type once common in Sussex which is usually the same front and back. It was put on over the head, unlike the coat-type of smock which could be unbuttoned
Ruddle pot	Tin or pot containing red chalk for marking ewes during tupping

Shearling A sheep after its first shearing during its second year of age

Sheep bells Used for locating sheep in poor weather or detecting a disturbance to the flock by their tone. They were prized possessions of shepherds, but only a selected number of sheep would wear them. Several types were used on the Downs including those made of iron such as the canister and cluckets; or those made of brass such as the cup, latten or crotal (rumbler) bells

Sheep bow Also known as blue bow, bilbo, strod or yoke. Forked shaped branch positioned in the ground for holding heads of sheep during dagging or trimming

Sheep counting A shepherd's particular way of counting sheep, varying from region to region

Sheep crooks The Downland Shepherd's badge of office. Used for catching sheep by the hindleg, or small lambs by the neck. Made of wrought iron, sometimes from a gun barrel, by the village blacksmith. The forges at Pyecombe, Falmer and Kingston by Lewes were renowned for their high level of craftsmanship. A Pyecombe crook cost 25s. during the 1920s.

Sheerman A man who comes from beyond the counties of Sussex and Kent. A term of derision

Shepherd's crown A fossil sea-urchin found in chalk. They were kept as charms (lucky stones) by rural folk

Shepherd's hut Originally a wooden structure on four wheels drawn by horses. Occupied by shepherds, primarily during lambing

Shepherd's shelter Hollow in bank with rough cover of furze or turf or a specifically cut thorn bush, giving protection from the weather

Ship Sheep

Snaps Shears. There was a shearing gang known as the Patcham Snaps

Smock Linen garments worn by shepherds and other rural workers as working clothes, which were sometimes weather-proofed with linseed oil. Working smocks were grey, black, or blue. White smocks were kept for best, being worn for marriage, funerals, and on Sundays. The precise use of this word in Sussex has been much disputed. Wills used it for both the round-frock and coat-type smock

Southdown sheep The Southdown is the oldest of the Down Breeds. For further details see Note 7, 'The Shepherd's Year'

Sundial Rough circle cut in turf, with appropriately positioned sticks. Used by shepherds for telling the time of day

Strap collar	Buckle strap with a little hole cut out to hold the top of a bell
Tackle	Fittings, including chin boards, lockyers, straps and yokes, used for fastening bells around sheep's neck
Tailing iron	Small spade shaped iron on rod with wooden or iron handle. Used for docking lamb's tails
Tar-boy	Boy member of shearing gang providing shearer with tar pot to dress wounds to sheeps' skin during shearing
Teg	Also known as hog or hogget. Young sheep before first shearing
Thatching needle	A pointed wooden stick with an eye or an iron needle used for securing thatch onto lambing fold hurdles
Trug	Wooden basket traditionally made at Hurstmonceux from willow with cleft ash or chestnut frame. Used by shepherds for carrying mangolds and turnips
Turnip peck	Tool with two prongs and long handle used for lifting turnips
Varmints	Vermin
Wattle	Woven hurdle made with split hazel. The gate hurdle is also referred to as a wattle by some shepherds
Wether	A castrated ram lamb
White Ram Night	Inaugural meeting of shearing gang in a public house to plan the season's work
Winder	Member of shearing gang allocated to roll up fleeces during shearing
Yeo	A ewe, pronounced 'yo' by the Downland shepherds
Yoke	(1) An arched piece of wood, also known as a crook, used for suspending sheep bells; (2) Also referred to as a sheep bow; a wooden triangle or suspended bar hung around a sheep's neck to prevent it escaping; (3) A wooden frame hung over oxen's neck, onto which harness is attached
Yolk	Grease in a fleece

— NOTES —

INTRODUCTION

1 In *Bypaths of Downland*, published in 1927.

2 Recorded in *Kelly's* Brighton directory for 1923.

3 C.E.M. Joad in his book of that title.

4 Charlie Yeates, of the rural life museum at Stanmer.

THE MAN AND THE HILLS

1 There was much interest in windmills during the 1920s and 30s, particularly in Sussex, which was famous for its mills. Barclay Wills' friend and fellow Downsman, Thurston Hopkins, was among several authors who wrote books about them.

2 Quite when Barclay Wills moved from London is unknown. He was living in Brighton in 1923 and 1924, and moved to Worthing soon after, probably in 1925.

3 The date of this first meeting is unknown, although it is likely that they knew each other well by the time of Barclay Wills' Christmas visit to Nelson in 1923. Barclay was to write more about Nelson than any other shepherd and the two remained close friends until Nelson's death at the age of eighty on 10 January 1943.

4 The South Downs and, more generally, Sussex were popular subjects for guide books and topographical writing before and after the First World War. There was intense interest also in the many archaeological features to be seen then on the (unploughed) Downs. Barclay's dry and gently humorous, 'Forget, if you can. . . .' may be seen as a good

natured rebuke to those whose obsession with such matters blinded them to the beauty of the Downs themselves.

5 Barclay Wills' term for this hobby of flint finding. While walking the Downs, Barclay usually kept a sharp eye out for flints of geological or archaeological interest, particularly during the winter months when ploughing had sometimes brought 'finds' to the surface.

6 Shepherd Fowler died in 1940. His crook was later given to East Dean Church in remembrance of an occasion when it was lent to a visiting bishop who was without his pastoral staff!

7 Red oxen, known as 'Sussex Reds'. They were the last team to work a Downland farm and belonged to Major Harding, of Birling Manor. Arthur Beckett described the team as 'recently revived' in 1927, but it seems to have been disbanded before the summer of 1929, the year this piece was first published.

8 It was still the custom of the older Downland shepherds to put bells on their sheep in the years after the First World War. Canisters were the type of bell most frequently heard on the Downs.

9 It is interesting that when Barclay Wills wrote this the mill was working, for two of its sweeps were badly damaged by a storm in November 1928. The mill still survives today.

10 Butterflies belonging to the *Lycaenidae* family, which also includes the Hairstreaks and Coppers. Five species of Blues are still found on the South Downs: the Small, Common, Holly, Adonis and Chalkhill Blues. The species of the co-operative female cannot be identified with certainty from Barclay Wills' description.

11   Barclay Wills was born in Islington. It is tempting to see 'the stranger' as a fictional alter-ego.

12   Waterhall Mill, to give it its local name, was the last windmill to be built in Sussex. Standing high on the Downs to the East of Patcham, the mill is a noted Downland landmark.

13   Barclay Wills was among the first to join the Society of Sussex Downsmen (in January 1924) and he remained active in it until 1939. As a District Officer of the Society, he made many reports concerning rights of way on the Downs near Worthing. The suggestion that he was 'mixed up with all the societies interested in archaeology' should be taken humorously. Although John Pull, one of his closest friends, was the leading excavator of the Worthing Archaeological Society from the 1920s onwards, Barclay was only briefly a member (1927–9) and seems thereafter to have been wary of such involvement. Surprisingly, he never joined the Sussex Archaeological Society.

14   Red earthenware made by rural potteries in Sussex from the late eighteenth until the early twentieth century. Barclay Wills is known to have had a substantial collection, now sadly dispersed.

THE SHEPHERDS OF THE DOWNS

1   The 1930s. 'The lure of the shepherd's work' was first published in January 1933 in the *Sussex County Magazine*.

2   Barclay Wills was collecting material for the *Shepherds of Sussex*.

3   Black oxen, also known as Welsh runts, which replaced the local 'Sussex reds' as draught animals in the latter half of the nineteenth century. By 1900 they were themselves an uncommon sight on the Downs, having been largely superseded by horses.

4   After Blann's death at Patching on 10 February 1934, Barclay Wills wrote a moving obituary which was published in *The Worthing Herald* on Saturday, 3 March.

5   Michael Blann's whistle pipe and song book may be seen at Worthing Museum, which has also published an edition of the song book. Other versions of some of the songs can be found in Bob Copper's books (see Further Reading) and have been recorded by the Copper family in *A Song for Every Season* (Leader Records, 1971).

6   Walter Wooler died on 23 April 1936, at Pyecombe. He was seventy-nine.

7   Common Mallow, *Malva sylvestris*. Widely used in folk medicine as a soothing agent.

8   Geoffrey Grigson in his *The Englishman's Flora* lists this as a local name in Somerset for Aaron's Rod, *Verbascum thapsus*, 'Medicinally given for coughs, gripes, piles.'

9   Not Garlic Mustard (*Alliara petiolata*), which was known in Sussex as 'Jack by the hedge', but Red Campion, *Melandrium rubrum*.

10   'A limited number of *brass* crooks were once made at Brighton. They were produced secretly at the railway workshop. The first one, made as a curio, resulted in requests for more, but many of those who acquired them found them liable to snap asunder during use. Very few specimens are to be found now, and those I have traced are kept as curios by their owners. My discovery of the first brass crook made is related in the interview with Charles Trigwell.' (*Shepherds of Sussex*, page 127.) Barclay Wills would seem to have found these specimens between 1927 and 1933 (when the passage just quoted was first published), as he stated in *Bypaths in Downland* that he had not seen one.

11   The road to Devil's Dyke, a prominent hill on the northern scarp of the Downs, near Fulking.

12   Chestnut was more usual for sheep wattles at this date, but was regarded as inferior by the older shepherds.

13   The ancient hill figure cut into the chalk slope of Wilmington Hill, and a famous Sussex landmark.

14   The Downland shepherds were popular as photographic subjects during the 1920s and '30s.

15   Barclay Wills does not reveal the shepherd's identity in this piece. A copy of *Downland Treasure*, however, survives with marginalia in Barclay's beautiful hand, a note on page 5 describing two partridge feathers stuck into the book as a: 'Souvenir of ramble with Nelson Coppard. Falmer 26/9/26'.

16   Nelson was then sixty-three years old.

THE SHEPHERD'S YEAR

1 Nelson Coppard.

2 The Ode, 'Intimations of Immortality from Recollections of Early Childhood', published in 1807. Another instance of Barclay Wills' gentle humour. Wordsworth's 'happy Shepherd-boy' was 'A six years' Darling of pigmy size!' Barclay was in his forties when he wrote this and, at about six feet, tall for his generation.

3 The ancient, sheep-cropped Downland sward has a soft, springy quality which 'gives' underfoot, making it delightful to walk on. Chalk grass-land which has been 'improved' for agriculture (by ploughing, re-seeding, chemical spraying, etc.) does not yield in this way and is much harder walking. Barclay Wills alludes to this difference elsewhere.

4 Still displayed in Worthing Museum.

5 This beautiful Downland butterfly was once abundant everywhere on the Sussex Downs between mid-July and the end of August. Only the male, whose upper wings are a pale, irridescent blue is 'glorious', the female's upper wings being dowdy brown.

6 The 1919 edition of Blaker's book, for which Barclay Wills' friend, Habberton Lulham, wrote the foreword.

7 Founded in 1892 and still in existence. The Southdown is the oldest of the Down breeds. It was developed by John Ellman of Glynde (1753–1832) during the late eighteenth century. It created much interest amongst royalty and the aristocracy keen to improve agriculture on their estates. The Duke of Richmond and the Earl of Egremont were two Sussex landowners who started flocks on their estates, at Goodwood and Petworth.

   The breed continued to be improved during the nineteenth century, notably by Jonas Webb of Babraham in Cambridgeshire. It became famous for the quality of its wool and mutton and was exported to many other countries, particularly New Zealand.

   Downland farming was based on a system of sheep and corn production. After the First World War, however, Southern sheep numbers declined rapidly – partly because preference was given to other breeds, partly because of the agricultural depression of the 1920s and 30s. The rams have been used extensively for crossing with other breeds, but only about 1,500 breeding ewes remain today.

8 The first Sussex newspaper, founded in the mid-eighteenth century, about which Arthur Beckett later wrote an article in his *Sussex County Magazine* (Vol. 15, 1941, pp. 247–54).

9 One of Barclay Wills' closest friends, Arthur Beckett, who died aged seventy-one in 1943, is today best remembered as the Founder Editor of *The Sussex County Magazine* and as the Foundation President of the Society of Sussex Downsmen.

10 'A Farthing in the Ear', *Sussex County Magazine*, vol. 4 (1930), p. 104. The article dates this meeting as 6 October 1929. It is particularly interesting as being the only entry to have been published from Barclay's diary, whose whereabouts is unknown.

11 Lloyd, who edited *The Southdown Sheep* (1933), corresponded with Barclay Wills during the 1930s and provided much of the material for the chapter on Sussex sheep in *Shepherds of Sussex*.

12 Identified by Barclay Wills as Frank Shepherd.

13 A very similar jug may be seen at Worthing Museum and may well have come from Barclay Wills' own collection. Oatmeal water, known as 'stokos', was provided rather than beer because the farmer, Martin Robinson, was a Quaker and so would not give the men alcohol.

14 Colin Andrews cites this as being among Michael Blann's repertoire, although it does not appear in his song book. It was among the songs that Bob Copper was later to sing to Barclay while visiting him.

15 Blann's song book has another version. See *Shepherd of the Downs* by Colin Andrews, p. 16.

16 Geering's book, first published in 1884 and titled *Our Parish: A medley by One Who has never lived out it*, was re-published in 1925 with an introduction by Beckett. Hollamby's version was called 'Gooche's Strong Beer' after the Hailsham brewer of that name.

17 Barclay Wills makes many references to this member of the Coppard family. Coppard was born at 8 Church Hill, Patcham, on 25 February 1887, and was christened Harry Charles, although he was later called Henry, rather than Harry. His father, Frederick Coppard, was also a shepherd. After the Second World War, Henry was one of the last of the Downland shepherds and, as a result, was much photographed. He was shepherd at Water Hall Farm and at Court Farm, where he worked when he retired at Michaelmas, 1951. Charlie Yeates, who

knew Henry well, wrote a moving tribute to him ('A Sussex Shepherd Retires', *Sussex County Magazine*, vol. 26, 1952, pp. 177–8), and after his death on 21 December 1963, *The Times* honoured him with a lengthy obituary ('Epitaph for a Downs Shepherd').

18 An isolated farm on the Downs near Sompting.

19 The largest Sheep Fair in Sussex held on Nepcote Green, the Fair has medieval origins and since 1796 has been held annually on the second Saturday in September. Barclay Wills doubtless made many visits to the Fair. This account was published in 1927. A newspaper picture shows Barclay standing between rows of sheep pens at the Fair in 1930. He wears a trilby and is carrying what appears to be a shepherd's bag.

Apart from Findon Great Fair, two other large sheep fairs were held at Lindfield and Lewes, and three lamb fairs at St. John's Common, Findon and Horsham. In addition there were the regular markets held for example at Chichester, Pulborough and Steyning.

20 1923: this article was first published in *The Downland Post* in December 1924. The canister bell was 5½ in high.

SHEPHERDS' GEAR

1 John Norris' glass-windowed lantern may still be seen there.

2 Tom Rusbridge had already retired from his work as a shepherd when Barclay Wills first met him in the 1920s. The two became firm friends, Barclay often calling at Tom's home at Nepcote, near Findon, while on a ramble. In an article about Tom in the *Sussex County Magazine*, (vol. 7, 1933, pp. 257–8), Barclay records that he 'was born on 12th July 1856, at West Firle, where his family had been for several generations.' Tom, who had begun work at seven, had a less settled life, working at a farm near Bosham; at Wick Farm, Ditchling; at Mary Farm, Falmer; at Hill Barn Farm, Southwick; and at Findon Park, his last farm.

He was a good musician and singer, and taught his youngest daughter, Grace, to play the melodian. Grace, now eighty-five, remembers that each year on the eve of Findon Sheep Fair shepherd friends of her father would meet at their home in Nepcote. While she

played her melodian, Tom and the others would sing the old Sussex songs. Among these shepherds she remembers Nelson Coppard, Walter Wooler, Michael Blann, George Humphrey, Charles Trigwell, and Richard Flint.

He died in 1932 and lies buried in Findon churchyard, near the Downs he so loved.

3 Wills is mistaken in attributing this quotation to the *Natural History of Selborne*. It comes from White's *Journal*, and is dated 18th January, 1769. It is likely that Wills used a Victorian edition of Selborne – such as Thomas Bell's of 1877 – which included material taken from the *Journal*.

DOWNLAND SHEEP CROOKS AND THEIR MAKERS

1 Barclay Wills later found examples of brass crooks, as referred to on p. 55.

2 Ash which has been cut back to ground level.

3 Nelson Coppard.

4 A record of a later visit to the Kingston forge is given in 'Two Sussex forges', *Sussex County Magazine*, vol. 4 (1930), pp. 74–6.

5 Horace Hoather, described by Barclay in his article, 'Two Sussex forges', ibid.

BELLS AND BELL MUSIC

1 This is described in *Downland Treasure*. 'A Discovery in London', pp. 158–61. Barclay Wills first learned of the Whitechapel Foundry from an article in the *Sussex County Magazine*, vol. 1 (1927), p. 414, by the Rev. A.A. Evans.

2 Dr. Edwin Percy Habberton Lulham (1865–1940). Dr. Lulham was a close friend of Barclay Wills, sharing many of his Downland interests, and contributed a chapter of 'Stray Memories' to *Shepherds of Sussex*. He was among the founding members of the Society of Sussex

Jerrome, P., and Newdick, J., *The men with laughter in their hearts*, Window Press, Petworth, 1986.

Jerrome, P., and Newdick, J., *Proud Petworth and Beyond*, Window Press, Petworth, 1981.

Jerrome, P., and Newdick, J., *Old and New . . . Teasing and True*, Window Press, Petworth, 1988.

Jesse, R.H.B., *A Survey of the Agriculture of Sussex*, Royal Agricultural Society of England, 1960.

Johnson, W., *Talks with Shepherds*, Routledge, London, 1925.

Keith, W.J., *The Rural Tradition*, University of Toronto Press, Toronto, 1975.

Lousley, J.E., *Wild Flowers of Chalk and Limestone*, Collins, London, 1950.

Massingham, H.J., *English Downland*, Batsford, London, 1936.

Pratt, C., *A History of the Butterflies and Moths of Sussex*, Booth Museum of Natural History, Brighton, 1981.

Pull, J.H., *The Flint Miners of Blackpatch*, Williams and Norgate, London, 1932.

Robinson, M., *A Southdown Farm in the Sixties*, Dent, London, 1938. Her memoir of life at Saddlescombe Farm in the 1860s is a classic of Sussex literature.

Tomalin, R., *W.H. Hudson: A Biography*. Faber and Faber, London, 1982.

Wales, A., *A Sussex Garland*, Countryside Books, 1986.

Walford, Lloyd E., *The Southdown Sheep*, Southdown Sheep Society, 1933.

White, G., *The Natural History of Selborne* (1789), Penguin Books, London, 1977.

Wills, Barclay, *Bypaths in Downland*, Methuen, London. 1927.

Wills, Barclay, *Downland Treasure*, Methuen, London, 1929

Wills, Barclay, *Shepherds of Sussex*, Skeffington, London, 1938.

Wolley-Dod, A.H., *Flora of Sussex* (1937), Chatford House, Bristol, 1970.

Young, the Rev A., *General View of the Agriculture of the County of Sussex* (1813), David and Charles, Newton Abbot, 1981.

Yeates, C., *Hovel in the Wood*, Privately published, 1986.

Readers who wish to hear an authentic rendering of Sussex folk songs, including some of those that Michael Blann wrote down in his song book, are referred to the Copper family's 'A Song for Every Season', Leader Records, 1971.

— Acknowledgements —

We would like to thank the very many people who have helped us in the compilation of this book by providing information, lending photographs and documents. We are most grateful to the following for their generous assistance:

The Armstrong Research Library and staff at the Weald and Downland Open Air Museum, Singleton: Mr. J. Mainwaring Baines: Mr. L.M. Bickerton: Mr. O. Buckle: Mr. L. Coverley: Mr. J.S. Creasy, Museum of English Rural Life, Reading University: Dr. P.G.M. Foster and Dr. M. Grainger of the West Sussex Institute of Higher Education: Mr. M. Hayes, Local Studies Librarian, and staff of the West Sussex County Library Service: Mr. G.A. Holleyman: Marian Hulland of the Worthing Natural History Society: Mrs. A. Induni of the Worthing Archaeological Society: Dr. G. Legg, Keeper of Biology at the Booth Museum of Natural History: Professor G. Lewis of Leicester University: Mr. J. Norwood, the Curator, and Dr. S. White of Worthing Museum and Art Gallery: Mr. C. Platt: Mr. J.M. Robinson, Librarian to the Duke of Norfolk: Mr. J. Roles, Keeper of History and Archaeology, Brighton Museum: Mr. P. Palmer: Mr. D. Prior, of the Royal Commission on Historical Manuscripts: Mr. D.G. Sutton of the Location Register of English Literary Manuscripts and Letters: Mr. Philip Palmer, Hon. Sec. of the Society of Sussex Downsmen.

In particular, we would like to thank those who knew Barclay Wills and his circle and who so generously gave of their time to talk to us about him, including Mrs. Heryet, Lyndon Mason and (the late) Charles Yeates, and Bob Copper who has so kindly written the foreword to this book. Finally, our thanks to Jane Pailthorpe for her typing and support, and to Mrs. D. Young for her work on the map.

Our text is taken largely from Barclay Wills' three published books: pages 1; 4; 6; 10; 11; 13–15; 45; 48–9; 51; 53; 68–70; 72; 91–2; 93; 96–7; 98–100; 121; 123; 125; 131–4; 135–6; 140–41; 144; 146–8; 154–6; 158 are from *Bypaths in Downland*; pages 6–8; 10; 40–42; 42–4;* 73; 114–17; 125–8; 128–30;* 134; 141; 148–54 are from *Downland Treasure*; pages 11–13; 16–17; 20–23; 25–6; 28–40; 53; 56–9; 61; 64–8; 73; 75–8; 87–8; 90; 93; 101–9; 112–14; 118–20; 123–4; 158–61 are from *Shepherds of Sussex*. (* Also reproduced in *Shepherds of Sussex*.)

For their generous permission to reproduce material from these works the editors and publisher would like to thank

— ACKNOWLEDGEMENTS —

Methuen & Co. Ltd. for the use of extracts from *Bypaths in Downland* and *Downland Treasure*, and Century Hutchinson Publishing Group Ltd. for those from *Shepherds of Sussex*. We should also like to thank Beckett Newspapers Ltd. for permission to reproduce part of the article, 'Two Sussex Forges', which was originally published in the *Sussex County Magazine*, and which appears in our text under the title of 'A Visit to Falmer Forge'. 'The Wheatear' and 'The Stonechat' first appeared, respectively, in the June and July issues of *The Downland Post*, 1924. Finally, we should like to acknowledge the kindness of The Society of Sussex Downsmen and of Mr. R.J. Huse, FLA, the County Librarian, West Sussex County Council for permission to quote from manuscripts in their possession.

PHOTOGRAPHS

We would also like to thank the following organizations for permission to reproduce photographs, or original art work: Beckett Newspapers Ltd., xviii; Mr. Bob Copper, 79, 145; East Sussex County Library Service, 145; Walter Langmead, 54, 55; The Mansell Collection, xxiv, 9, 14, 52, 82, 157, Endpapers, Back Jacket; West Sussex County Library Service, iii, vii, xiii, xiv, 5, 7, 27, 132, 137, 138, 139; West Sussex County Record Office (The Garland Collection), 50, 71, 153, 164.

While every effort has been made to contact all copyright holders, in some cases we have been unable to trace them, for which we apologise.

Index

Alciston, 38
Aldbourne (Wilts), ix, 108
Alfriston, 31, 88, 107
Arnold, 'Stumpy', *126*, 127
Ashington, 149, 150
Ashurst Mill, 148–50, *149*
Allcock, Albert, 88, *106*, 107

Bailey, George, 39–40, *39*, 57
Bately, Mrs. Lillian, xvi, xx
Beachy Head, 8
Beckett, Arthur, xvii, xix, xx, 58, 67; *Spirit of the Downs, The*, xvii, xviii; *Sussex County Magazine*, xvii, xix, 58, 172, 173, 175
Beecher, John, *119*, 120
Beeding, 26
Beeding Court Farm, 39, 57
Belloc, Hilaire, *The Four Men*, xvii, xviii
Bells: Cattle, 48; *See* Sheep Bells and Tackle
Berry (Blacksmith, Pyecombe), 37, 70, 98–9
Bickerton, L.M., xxiii
Birds: Barn Owl, 158, 176; Corncrake, 22–3, 11–12; Cuckoo, 11, 121; Dartford Warbler, xiii, xv, 136, *137*, 138; Goldcrest *vii*; Goldfinch, 135, 140, 144, 148, 156; Kingfisher, 6, 7; Landrail (*see* Corncrake); Linnet, 4, 51, 121, 131, 133; Magpie, 43, 121, 124, 132, 133, 155; Nightingale, 6; Nuthatch, xii; Partridges, 43, 121, 124, 159; Peewits, 132, 159; Rook, 12, 121, 129, 148, 158, 159; Shufflewing, 135; Skylark, 131, 132, 140, 144; Sparrow hawk, 8; Stonechat, xii, 42, 49, 133, 135, *138*; Tawny Owl, 121, 124; Tree Creeper, xii, *xiii*, 51, 135; Wheatear, xi, 8, 51, 131, 133, *139*; Winchat, 156; Wren, xiii, 131, 135–6, *137*; Yellow Hammer, 131, 133, 135, 136
Blackmore, Stephen (Shepherd), xvii
Blackpatch Hill, xix, xx, xxi, 175
Black Ram Night, 59, 64
Blaker, Nathaniel, 172; *Sussex in Bygone Days*, 56
Blann, Michael, 23, 25, 26–9, *27*, 32, 67, 79, 83, 87, *89*, 90, 114; Music, 28, 32, *89*, 90
Brighton, xiv, xv, 10, 34; Crooks, 92; Museum, xxi, 97
Browns Farm (West Blatchington), 35, 36, 38, 98
Browns Farm (Rottingdean), xvi
Butterflies: Adonis Blue, 133, 171, 175; Blues, 131, 132, 171; Chalkhill Blue, 49, 171, 173; Clouded Yellow, 1,

141, 175; Green Hairstreak, 140–1, *142*, 143, 175; Peacock, xii, 156; Silver Studded Blue, xv; Small Copper, xxi, 10; White Admiral, xi

Captain (Shearing Gang), 61, 64
Chanctonbury, *111*, 150
Chant, George, *164*
Cissbury, *2*, 11
Cissbury Ring, xvi, xx, 120, 131–2, 140, 156, 175
Claspknife, 77
Clayton Mills, *xxiv*, 21, 127–8
Colts, 35, 61, 66
Coppard Frederick, 173
Coppard Henry (Harry), 69, 106, 144, 173–4
Coppard, Nelson, xv, xvi, 1, 4, 21–3, 42–4, 51–3, 56, 69, 78, 87, 93, 96, 108, *122*, 123, 125, 144, 171, 172, 173, 174, 175
Copper, Bob, ix, x, xx; *Song for Every Season, A* (record), 172
Cornfields on the Downs, 10, 134
Costrel, *157*
Counting rhymes, 59
Cox, Jack, 23–6, *24*
Cream Spot Tiger Moth, *xiv*
Cuckmere Valley, 11
Curwen, Dr. E., *Archaeology of Sussex*, xx

Devils Dyke, 78, 112, 116, 120
Dewponds, 111, 121–4, *122*, 133, 153
Dipping Hook (*see* Hook)
Dirt Knocker, 75, 76, 78
Ditchling, 36, 133; Beacon, 11, 14
Dormouse, *5*, 6
Downland Bottom, 13, 51, 102, 133
Downland Footpaths, 13, 66
Downland Turf, xi, xv, 20, 43, 45, 51, 101
Downland Post, The, 174

Durrington, xv, xvi, 175
  Coate Farm, 81
Dyke Hill, 21, 108, 112

Eastbourne, 8
East Dean (East Sussex), xvii, 6, 7, 10, 38; Birling Farm, 7, 146
Egerton, Revd J.C., *Sussex Folk and Sussex Ways*, 83–4
Ellacombe, Revd H.T., *Bells of the Church*, 102, 103
Ellman, John, 173
Exceat, 6, 144, 176
Evans, Revd A.A., 174

Falmer, xiii, 1, 10, 11, 45, 78, 108, 116, 140; Crookmaking, 97–8; Mary Farm, xv, 1, 21, 144
False Tongues, 38, 83
Farthing Mark, 31, 58
Findon, xiii, 158, 174; Church Hill, xx; Fair, 26, 33, 34, *62*, 69–72, *71*, 88, 98, 158, 160, 174; Mount Carey, xx, xxi; Nepcote Green, 174
Flail, 150, 151
Flint, Richard (Dick), 84, 98, 174
Flowers: Bee Orchid, 99; Birds Foot Trefoil, 132; Burnet Rose, 10, 11, 126; Clustered Bell Flower, 67; Cornflowers, 134; Cowslip, 144; Devils Bit Scabious, 11, 132; Dogs Mercury, 49, 129; Harebell, 126, 141; Heather, 133; Jack-in-the-Hedge (*see* Red Campion); Mallow, 32, 172; Ox-eye daisy, 126; Poppy, 10, 125, 134; Pride of Sussex (*see* Round-headed Rampion); Red Campion, 53, 172; Round-headed Rampion, 11, 101, 126, 141, 155, 176; Scabious, 11, 126, 141; Speedwell, 129, 141; Star of Bethlehem, 78; Thyme, 13, 132, 133, 140; Vipers Bugloss, 6; Violet, 51; Wild Arum, 49
Fly powder tin, 77
Folk Songs, 28, 53, 64, 66–8
Fowler, Dick, 6, 7, 8, 106, 146, 171
Fowler, Jim, xvii, *9*, *157*
Foxhole, 6
Frail, *9*, *14*, *74*, 77
Frost, Miss Marian, xvii
Fulking, 56–7, 172
Funnell, Charles, 40–2, *41*, 106–7, *106*
Furriner, 154–5
Furze, 11, 42, 43, 45, 49, 51, 109, *132*, 133, 138, 140

References to illustrations are in italic

Garlic Mustard, 172
Geering, Thomas, *Our Sussex Parish*, 67, 173
Gerard, Miss Ethel, xvii
Goad, *145*, 146
Golf, xv, 155–8, 132
Goring, 76, 80
Gorringe, E.J., 144, 176
Gorringe, Albert, 13, 116–7, *116*
Gorringe, Solomon, 116–7
Gosset, Miss A.J.L., *Shepherds of Britain*, xvii, 175
Green (Blacksmith, Falmer), 97–8
Grigson, Geoffrey, *Englishman's Flora, The*, 172

Hampstead Heath, xii
Harding, Major W.E., 146, 171
Harrow Hill, 25, 175
Hazelgrove, Jack, 65–6
Highgate, xii, xiv
Hoare, James, xii
Hoather, Horace, 174
Hoather (Blacksmith, Kingston), 93, 96–7, *97*, 144
Hollamby, John, 67, 173
Hook (*see* Sheep Crooks); Dipping, 22, *55*, 78
Hopkins, R. Thurston, xvii, 171; *Sussex Pilgrimages*, xviii
Horn Cup, xvi, 64
Humphrey, George, xv, 32–3, 58, 68–9, *68*, *79*, 88, 174
Hudson, W.A., *Nature in Downland, A Shepherds' Life*, xv, xvii, 136
Hurdles, 26, 51, 58, 70, 129; Making, 146–8, *147*; Wattles, 37, 155

Isle of Wight, xii
Islington, xii, 171

Jefferies, Richard, *Wildlife in a Southern County, Nature Near London*, xvii, xix
Joad, C.E. of, xvi, 171

Kingston-by-Lewes, 10, 144; Crookmaking, 8, 92, 93, 96–7, 174
Kipling, Rudyard, *Puck of Pooks Hill*, xviii
Knobbing iron, 96, 144

Lambing, 45–53, 81, 146, 155; Bottlefeeding, 45, 49, *52*, 53, 75; Fold, *18*, 19, 48, *50*, 75, 129, 135, 155
Lamb(s): Creep, 76; Hob, 45; Jacket, 21, *22*

Lewes, 11, 38, 39–40; Bells, 106, 107; Fair, 28, 30, 36, 103, 116, 166
Lieutenant (Shearing Gang), 61
Lockyers, *see* Tackle
Lucas, E.V., xviii
Lulham, Habberton Dr. E.P., xvii, 103, 173, 174–5

Marking Iron, 57
Martin, E.A., 123, 175
Massingham, H.J., *English Downland, Shepherds Country*, xvii, xix
Mitchell, Charles (Blacksmith, Pyecombe), 25, 31, 99–100
Moulding, Jesse, 33, *33*, 64, 80

New Forest, xii, 133
Newell, G., *77*, 78, 108, 112, 120
Norris, John, 81, 174

Oxen, 7, 23, 37, 99, 144–6, *145*, 150–1
Ox-ploughing, 146; Shoeing, 42, *145*, 150; Shoes, 12, 23, 43, 77, 96, 99; Yoke, 7, 144

Paddock, bertha (*see* Wills, Bertha)
Paddock, James, xiii
Patcham, 21, 36, 69, 106, 150, 172; Mill, 11, 172; Water Hall Farm, 144, 173
Patching, xv, 25, 26, 65, 172
Peacehaven, xvi
Pitching iron, 76
Poynings, 11, 21
Price, Miss Nancy, xxii–xxiii
Pull, John, xvii, xx, xxi, xxiii, 172, 175; *Flint Miners of Blackpatch*, xx
Pyecombe, 21, 30, 31, 58, 92, 127; Crookmaking, 25, 37, 70, *92*, 98–100; Forge, 31, 99, 100

Rabbit Catching, 43–4, 126–7, *126*, 153
Rape greens, 34
Rewell, Henry, *14*, 22
Richardson, Mr., 150
Robinson, Maude, xvii, 60
Roller-Wattle, 85–7, *86*, *168*
Rottingdean, 36 (*see also* Brown's Farm)
Ruddle pot, 77
Rubbing rail, 87
Rusbridge, Tom, xv, xvii, 14, 83, *83*, 84, 174

Sainsbury, C.E., xix–xx
Saddlescombe Farm, 21, *60*, 61
Salvington Mill, xxii, 140; Ridge, xvi
Shearing Gangs, 39–40, 59–62, 64–6
Shears pouch, *74*, 76, 159
Sheep Bells, xxii, 101–117, *104*, *106*; Canister, 10, 25, 28, 53, 99, 103, 104–5, 107, 133, 151; Ring of Canisters, 7, 38, 105, 114–7; Cluck, cluckets, 48, 103, 105, 123, 133, 156; Crotal, 108, 112; Cup, 107, 112; Horse, 8, 107, 112; Latten, 11, 28, 48, 107; Lewes, 106, 107; Music, 20, 101–2, 108, 133; Rumbler, 48, 108; Sale of, 103; Use of, 101–2; White Latten, 107–8; Wide mouthed brass, 7, 40, 106–7; Wide mouthed iron, 105
Sheep Bows, 68–9, *68*, 170
Sheep Cages, *50*, 148
Sheep Counting, 58–9
Sheep Crooks, 8, 70, 91–100, *92*; Brass, 313, 34, 92, 172, 174; Brighton, 92; Greens, 92; Handles, 73; Kingston-by-Lewes, 8, 92, 93, *97*; Pyecombe, 25, 37, 70, 98–100; Sticks, 93, 96; Wooden, 109
Sheep Cures, 31, 32, 75
Sheep Dogs, 23, 25–6, 35, 38, 42, 48, 118–20, *119*, 123, 133, 175
Sheep Fairs (*see also* Findon), 36–7, 39
Sheep Marking, 22, 30, 31, 57–8, 76
Sheep Names, 48
Sheep Prices, 69
Sheep Shears, 29, 69, 73, 76
Sheep Shearing, 59–61, *60*, 64–6
Sheep Trimming, 68–9, *68*
Sheep Washing, 53, *54*, *55*, 56–7
Sheerman (*see also* Furriner), 155
Shepherd, Frank, 65, 173
Shepherd, William (Old Shep), *153*
Shepherds Bush, 31, 32
Shepherd's Clothes, 23, 25, 26, 28, 31, 33, 38, 82, 82–5
Shepherd's Crown, 12, 13, 75, 77, 136, 141
Shepherd's Gear, *46*, 73–90, *74*
Shepherd's Hats, 23, 26, 34, 38, 82, 84
Shepherds Hut, 45, 47, 49, 51, 73, 75, *94*, 95, 130
Shepherd's Songbook and Whistle-Pipe, 28, 29, 88, *89*, 90
Shepherd's Songs, 64, 66–8
Shepherd's Sundial, 23, 87–8

Shepherd's Stool, *77*, 78
Shepherd's Umbrella, 40, 78–80, *79*, 85, 148
Shepherd's Wages, 26, 82
Smocks, 26, 28, 31, 33, 34, 40, 78, 82, 84–5
Society of Sussex Downsmen, xvii, xvi, 172
Sompting, 26, 70, 114
Southdown Sheep, 56, 57, 58, 129, 146, 151, *161*, 173
Southdown Sheep Society, 57
Stanmer Down Pond, 123, 124
Standean, 36
Steyning, 23
Sturt, George, xv
Sussex: Literature, xvii–xviii; Plough, 150, 151; Pottery, xvi, 15, 64, 172, 173; Mills, 151, 171; Museum, 151
Sussex Archaeological Society, 172
Sussex County Magazine, xvii, xviii, xix, 58, 172, 173, 174, 175
Sussex Weekly Advertiser, 58

Tackle, 28, 109, 112–4; Bell Straps, 113; Chin Boards, 108, 109, 112, 114; Lockyers, 28, 113–4, *113*; Pegs, 109, 112, 113, 114; Readers, 113; Staples, 113; Strap Collars, 112; Tongues, 108; Yokes, 28, 109, 112, 113, 114
Tailing Iron, 76
Tally Sticks, 59, 96
Tarboy, 61, *65*, 66
Tegboy, 17, 21
Tegs, 90, 160
Thatching Needle, 12, 75, 78
Thorburn, Archibald, xiii, xiv
Toms, Herbert S., xvii, xxi
Trigwell, Charles, 33–5, *33*, 172, 174

Trigwell, Tom, 34, 35
Trug, 42, 75, 130
Turner, Shepherd, *82*
Turnip pecker, 76

Upton, Frank, 35–7, *36*, 84
Upton, Walter, 98

Walford, Lloyd, H., 59, 173
Washington, 75, 150; Lee Farm, 23, 25
Wattles, 37, *71*, 155, 172
Weald, The, 148
Webb, Jonas, 173
Wells, Robert, 107, 108
West Blatchington, 117 (*see also* Brown's Farm)
West Dean (East Sussex), 6, 146
West Firle, 38, 174
Whitechapel Bell Foundry, 103, 108
White, Gilbert, 87; *Natural History of Selborne*, 85
White Ram night, 61
Wills, Barclay: Life – appearance, *viii*, ix, x, 4, 173; health, ix, xx, 13; London years – birth, parents, xii; early interest in the countryside, xii, 1; employments as clerk, xii; marriage, xiii; birth of daughter, xiv; move to Brighton, xiv; business difficulties, xiv–xv; move to Worthing and Durrington, xiv–xv; Worthing and war years, xv–xxiii; death and posthumous exhibition, xxiii.
Interests and Views – archaeology and flinting, xix–xx, 12, 126–7, 132, 171, 172; bird watching, 131–3; 135–40; butterflies, xxi, 1, 10, 140–1, 175, 171, 173; bygones, collections of; shepherding tackle, x, xvi, 1, 4, 7, 8, 15, 172, 173; Sussex pottery, xvi, 172, 173; anger at destruction of Downland, xxi; conservation issues, xvi; contributions to Worthing Museum, xvi,

172–4; plea for a Sussex Museum, 150–1; Den, the, x, xx; dislikes; airplanes on Downs, 127–8; golf, 155–8; pylons, xxi.
Membership of Societies – Rambling, xv, 4–12, 131–4, 148–50; sketching, xii–xiii; 1; Sussex Downsmen, xvi, 172; Worthing Archaeological Society, 172.
Friendships – Arthur Beckett, xvii, xix, xx, 173; Nelson Coppard, xv, 1, 21, 22, 23, 42–44, 171; Bob Copper, ix, x, xx, 173; Habberton Lulham, xvii, xx, 173, 174–5; Nancy Price, xxii–xxiii.
Writings – *Bypaths in Downland*, xi, xviii, xix, 171, 172, 175; *Discoveries in Downland*, xx, xxi, xxii; *Downland Treasure*, xi, xii, xviii, xix, 172, 174; *Shepherds of Sussex*, xii, xv, xvi, xix, 172, 173, 174.
    Articles in the *Sussex County Magazine*, 172, 173.
    Articles other sources, 172, 174.
Sketches – vii, xii, xiii, 5, 7.
Wills (Family): Bertha (wife), xiv, xx, xxii; Henry William (father), xii; Henry William (grandfather), xii; Lydia (mother), xii; Mollie Barclay (daughter), xiv, xx, xxii; Sophia (grandmother), xii
Wilmington, 40, 114, 150, 172
Wotstonbury, 11, 99
Wooler, Walter, *30*, 30–1, 34, 57, 58, 99, 100, 172, 174
Wooler, W.E., 145, 146
Wordsworth, William, xxii, 45, 173
Worthing, xi, xiv, xv
Worthing Archaeological Society, 172
Worthing Herald, 172
Worthing Museum, xvii, xxiii, 42, 75, 81, 105, 172, 173

Yeates, Charles, xx, 171, 173
Yokes: Bell, 28, 93, 109, 112; Ox, 7; Sheepbow, *68*, 70